The Anatomy of Inclusive Cities

Creating cities inclusive of immigrants in Southern Africa is both a balancing act and a protracted process that requires positive attitudes informed by accommodative institutional frameworks.

This book revolves around two key contemporary issues that cities around the globe are trying to achieve – viz. the need to build inclusive cities and the need to accommodate immigrants. The search for building inclusive cities is an on-going challenge which most cities are grappling with. This challenge is complicated by the need to include immigrants who are always side-lined by policies of host countries. This book discusses the host–immigrant interface by providing a detailed insight of anchors of inclusive cities and a holistic picture of who immigrants are. These are then discussed contextually within the Southern African region, where insight into selected cities is provided to some depth using empirical evidence.

The discussion on inclusive cities and immigrants is a universal narrative targeting practitioners and students in town and regional planning, urban studies, urban politics, migration and international relations. The Southern African region once more provides an opportunity to further interrogate and understand the dynamics of immigration in selected cities. This book will also be of interest to policy makers dealing with challenges of inclusivity in the light of immigrants.

Hangwelani Hope Magidimisha-Chipungu is Professor of Town and Regional Planning at the University of KwaZulu-Natal. She is also a SARChI chair for inclusive cities supported by the National Research Foundation and South Africa City Networks. As a practitioner, she is a member of the National Planning Commission appointed by the President of the Republic of South Africa. Professor Magidimisha-Chipungu is the Academic Leader in the Planning and Housing Discipline at the University of Kwazulu-Natal and a board member of the South Africa Council of Planners.

Lovemore Chipungu holds a PhD in Town and Regional Planning, a master's degree in Rural and Urban Planning as well as a bachelor of science degree (honours) in Rural and Urban Planning. He is an active member of the Zimbabwe Institute of Regional and Urban Planners as well as a Corporate Member of the South African Planning Institute. He has published widely in peer-reviewed journals and books, and his areas of interest are in housing policy, land-use planning and urban design.

Routledge Studies in Urbanism and the City

Marketplaces
Movements, Representations and Practices
Edited by Ceren Sezer and Rianne van Melik

Designing Healthy and Liveable Cities
Creating Sustainable Urban Regeneration
Marichela Sepe

The Political Economy of Land
Rent, Financialization and Resistance
Edited by Mika Hyötyläinen and Robert Beauregard

Future of the City Centre
Global Perspectives
Edited by Bob Giddings and Robert J Rogerson

Political Economy of Housing in Chile
Francisco Vergara-Perucich, Carlos Aguirre-Nuñez, Felipe Encinas, Rodrigo Hidalgo-Dattwyler, Ricardo Truffello and Felipe Ladrón de Guevara

Lahore in the 21st century
The Functioning and Development of a Megacity in the Global South
Mohammad A. Qadeer

Writing the City Square
On the history and the histories of city squares
Martin Zerlang

The Anatomy of Inclusive Cities
Insight into Migrants in Selected Capital Cities of Southern Africa
Hangwelani Hope Magidimisha-Chipungu and Lovemore Chipungu

For more information about this series, please visit www.routledge.com/Routledge-Studies-in-Urbanism-and-the-City/book-series/RSUC

The Anatomy of Inclusive Cities
Insight into Migrants in Selected Capital Cities of Southern Africa

Hangwelani Hope Magidimisha-Chipungu and Lovemore Chipungu

LONDON AND NEW YORK

First published 2023
by Routledge
4 Park Square, Milton Park, Abingdon, Oxon OX14 4RN

and by Routledge
605 Third Avenue, New York, NY 10158

Routledge is an imprint of the Taylor & Francis Group, an informa business

© 2023 Hangwelani Hope Magidimisha-Chipungu and Lovemore Chipungu

The right of Hangwelani Hope Magidimisha-Chipungu and Lovemore Chipungu to be identified as authors of this work has been asserted in accordance with sections 77 and 78 of the Copyright, Designs and Patents Act 1988.

All rights reserved. No part of this book may be reprinted or reproduced or utilised in any form or by any electronic, mechanical, or other means, now known or hereafter invented, including photocopying and recording, or in any information storage or retrieval system, without permission in writing from the publishers.

Trademark notice: Product or corporate names may be trademarks or registered trademarks, and are used only for identification and explanation without intent to infringe.

British Library Cataloguing-in-Publication Data
A catalogue record for this book is available from the British Library

Library of Congress Cataloging-in-Publication Data
Names: Magidimisha-Chipungu, Hope, 1984- author. | Chipungu, Lovemore, author.
Title: The anatomy of inclusive cities : insight into migrants in selected capital cities of Southern Africa / Hangwelani Hope Magidimisha-Chipungu and Lovemore Chipungu.
Description: Abingdon, Oxon ; New York, NY : Routledge, [2023] | Series: Routledge studies in urbanism and the city | Includes bibliographical references and index.
Identifiers: LCCN 2022053069 (print) | LCCN 2022053070 (ebook) | ISBN 9781032026640 (hardback) | ISBN 9781032026657 (paperback) | ISBN 9781003184508 (ebook)
Subjects: LCSH: Africa, Southern--Emigration and immigration. | Immigrants--Africa, Southern--Social conditions. | Social integration--Africa, Southern. | Africa, Southern--Politics and government.
Classification: LCC JV9006 .M34 2023 (print) | LCC JV9006 (ebook) | DDC 325.68--dc23/eng/20230106
LC record available at https://lccn.loc.gov/2022053069
LC ebook record available at https://lccn.loc.gov/2022053070

ISBN: 978-1-032-02664-0 (hbk)
ISBN: 978-1-032-02665-7 (pbk)
ISBN: 978-1-003-18450-8 (ebk)

DOI: 10.4324/9781003184508

Typeset in Times New Roman
by SPi Technologies India Pvt Ltd (Straive)

Contents

List of Figures vii
List of Tables viii
List of Boxes ix
Acknowledgement x

1 Setting the Agenda for Inclusive Cities 1

2 Research Methodology 20

3 Immigrants and the City: A Conceptual Framework 41

4 The Southern African Region in a Historical Perspective 50

5 Unpacking Migrant Laws in Gaborone: 'The Gem of Africa' 64

6 Malawi: A Retreat into Lilongwe the 'City Centre' 80

7 Maputo: The Lush Capital of Mozambique 98

8 eGoli: Beyond the Splendour of Johannesburg through the Eyes of Migrants 114

9 Lusaka: A Retreat into Zambia 136

10 In the Shadows of the Sunshine City of Zimbabwe: Harare 153

11 Inclusion of Foreign Migrants in South African
    Cities: An Aside   172

12 Reflections on Immigration Policies and Inclusivity
    in Southern African Cities   193

13 The Epilogue   203

    Index   214

Figures

2.1	Data Analysis	28
5.1	Map of Botswana Showing the Position of Gaborone	65
6.1	Map of Malawi Showing the Position of Lilongwe	81
7.1	Map of Mozambique Showing the Position of Maputo	99
8.1	Map of South Africa Showing the Position of Johannesburg	115
9.1	Map of Zambia Showing the Position of Lusaka	136
10.1	Map of Zimbabwe Showing the Position of Harare	154
11.1	Spatial Integration in South African Communities	173
11.2	Level of Comfort in Local Residences	175
11.3	Restrictions on Movement	175
11.4	Rating Cities on Interventions for Improving Community Integration	176
11.5	Freedom of Access to Public Spaces	178
11.6	Fairness in Service Delivery	179
11.7	Freedom of Movement	179
11.8	Word Cloud Propositions for Improving Spatial Integration	179
11.9	Trust of Law Enforcement Agents	180
11.10	Security Concerns of Foreign Migrants	182
11.11	(a) Employment Sector, (b) Employing Institutions, (c) Employment Status	184
11.12	Employment Challenges	184
11.13	Economic Integration	185
11.14	Indicators of Access and Security in the Economic Sector	185
11.15	Awareness of Institutions Supporting Migrants	186
11.16	Home Affairs and Local Governance	187
11.17	Foreigners' Views on Xenophobia	188
11.18	Inclusive across Sectors	189

Tables

1.1	Sustainable Development Goals and the Urban Agenda	5
1.2	The Many Definitions of Inclusive Cities	6
1.3	Anchors of an Inclusive City	7
1.4	Typology of Immigrants	9
1.5	Economic Benefits of Immigrants	11
3.1	Levels of Social Analysis	44
4.1	Outcome of the *Mfecane*	51
4.2	Migrant Labour from Botswana, Lesotho and Swaziland to South Africa	54
4.3	Labour Agreements	56
4.4	Countries in the SADC Region	58
4.5	Immigrants in the Region	59
11.1	Level of Integration and Hostility towards Foreign Migrants	174
11.2	Rating Comfort Levels across Municipalities	174
11.3	Limitations or Restrictions on Your Movement in the City	176
11.4	Spatial Integration into Local Communities	177
11.5	Correlation Matrix – Spatial Integration	181
11.6	Correlation Matrix – Security and Safety	182
13.1	Rights of Immigrants	207

Boxes

1.1	Dimensions of Inclusivity in Cities	6
3.1	The Spatial Impact of Migration and International Economic System on the Inclusivity of Cities	43
4.1	Impact of the Pre-colonial Migrations	52
4.2	Historical Insight into Immigration Dynamics and the City	60

Acknowledgement

The discussion on immigrants and the city stems from the search for inclusivity in contemporary cities. In the context of this book, it was born out of the call by the National Research Foundation (NRF) of South Africa and the South African Cities Network (SACN) to interrogate inclusivity in cities. This, in turn, provided an opportunity to map and investigate the extent to which inclusivity of cities accommodates the vulnerable members of society (i.e. women, children, the elderly, migrants and those living with disability). The authors, therefore, extend their gratitude to the unwavering support of these institutions for funding this research. Together with the University of KwaZulu-Natal, they created an enabling environment through which the research and publication of this book were made possible. Above all, the unwavering support of our research team, which committed its valuable time in collecting and analysing data cannot go unnoticed. Indeed, it is through this unselfish support that our book became a reality. Lastly, we are grateful to our children – Tamai, Kundai and Tariro – who had to endure long hours of solitude while we were busy travelling across Southern Africa in order to decode the anatomy of inclusive cities.

1 Setting the Agenda for Inclusive Cities

1.1 Introduction

The emergence of the discourse on inclusivity in recent years is gradually changing our perception of equity and integration in contemporary cities. This evolves out of the realisation that cities, despite being centres endowed with opportunities, are also centres of attrition, largely borne out of diverse individual and group interests. At the centre of these conflicts are social, economic and political dynamics which emerge as strong circles of contention, ring-fenced by institutional mandates crafted by those in power. In the ensuing process, the voices of the vulnerable are muffled and their existence relegated to oblivion through marginalisation. Cities, in the meantime, emerge as mere physical artefacts, screened from 'societal haemorrhage' born out of socio-economic and political inequities. Yet, despite these negativities, urbanisation, through immigration and other avenues, continues to change these cities demographically and spatially. Under such prevailing circumstances, the search for inclusivity in cities is not by accident, but a genuine intervention bent on building equity in urban societies.

This chapter sets the agenda for inclusive cities with immigrants in mind. It stems out of the premise that among other factors that contribute to the growth of cities are immigrants. In reality, local, regional and international immigrants equally contribute to this natural phenomenon of urbanisation. However, the society, governed by various institutional frameworks, responds differently when it comes to the reception and assimilation of these immigrants. While most policies respond favourably to local immigrants due to their national status, international immigrants are subjected to a plethora of regulations in order to qualify and be accepted in the host society. Depending on their immigration status (be it legal or illegal; documented or undocumented), their existence is riddled with challenges which make their survival unbearable. Hence, immigrants emerge as one group of people that is vulnerable in urban areas. The presence of the elderly, disabled, women and children aggravates the conditions of these immigrants in the face of subtle as well as open hostilities emanating from the host society. It is under such situations that the search for inclusivity in cities translates into a necessary need bent to create harmonious societies irrespective of diverse backgrounds.

DOI: 10.4324/9781003184508-1

Within this same space of resonance, Southern African cities are not spared of the spate of immigrants. Most of these cities, which can be qualified as 'European Cities', emerged at the peak of capitalism – colonialism – which saw international capital being used in grand international projects. Hence, most major cities of Southern Africa were born out of the fusion of international capital and cheap migrant labour which was used to promote further accumulation and investment into a capitalist social and physical construct. In reality, migration and the resultant immigrants ran into the historical 'lifeline' of these cities. This situation was further complicated by the search for freedom by nationalist movements which, in the process, generated more immigrants in the form of refugees, as civilians caught up in wars sought asylum. Hence, the presence of immigrants in cities of Southern Africa is not by accident but forms part of a necessary capitalist production system which permeates society through various socio-economic and political dynamics. As such, the search for inclusivity in these cities will continue in a bid to create communities that are responsive to human cause, irrespective of accentuating circumstances that led to the status quo.

1.2 The Anatomy of Inclusivity in Cities

As we embark on the journey in search for inclusivity of immigrants in cities, irrespective of factors that necessitated their situation, it is necessary to reflect on this concept of 'Inclusivity in Cities'. Generally, the concept of inclusive cities emerges as an omnibus term which anchors on the need for equity to all urban residents. While this sounds very enticing and exciting, there is still need to dissect this terminology into finer details to enhance not only comprehension but also functionality of the urban environment vis-à-vis the urban population. On that note, inclusivity in cities is a function of multiple and diverse factors that bring different dimensions to our understanding of the city.

1.2.1 The Right to the City

The preamble to providing a platform for democracy and unfettered participation in cities can be borrowed from Lefebvre's 'Right to the city' notion coined in 1968. Lefebvre's right to the city encapsulated a radical approach that rejects the logic of capital and the dictates of the state control mechanisms (Lefebvre, 1996, Lefebvre, 2003). This, as Harvey (2012) contends, requires complete restructuring of social rights in order to facilitate the human rights to access, use, enjoy and participate in the city's production space.

The spatial component broadens the conceptualisation of space as more than just physical, but also as a force that governs justice, injustice and social, environmental and political processes that shape human life (Soja, 2010). This is further espoused by Soja (2010) who explains it from a spatial justice perspective by arguing that the right to the city is based on 'fair and equitable distribution in space of socially valued resources and opportunities to use them'. Hence, the right to the city focuses on what 'individuals believe is their spatial right versus the rights of others' (Duke, 2010). As Duke (2010) further explains, the notion depends on how homogeneous or diverse society is; whether public space is appropriate for all society within it and whether society (or individuals or groups) are able to inform and influence decisions on the public urban space within which they operate.

The sociological concept describes the right of individuals to belong to and influence the urban spaces that form part of their daily life and the rights they have within this notion of social, political, economic, educational, health, recreation and urban accommodation platforms. The notion encapsulates the right to participate in quality urban life. Lefebvre's concept was a response to social movements against undemocratic governments of exclusion in Europe, where the need was to remove 'the forces of alienation … active in urban space' (Aalbers and Gibb, 2014). This speaks against the privatisation and commoditisation of space to urban space having use value over capitalist value. It should be noted that the *right to the city* does not typify a city but encapsulates the 'right to belonging to any place' (Aalbers and Gibb, 2014). This is further emphasised by Soja (2010) who makes a categorical observation about Los Angeles' capitalistic production of space at the detriment of social space since it creates exclusion in the design and use of space in order to exclude unwanted populations. The results are high levels of class stratification which are not only depictions of disparity but also a display of power through control of space (Harvey, 2012). Harvey (2012) reiterates Lefebvre's observation in the realisation that those with capital influence state decisions on development and should not be the sole decision makers because they make capitalist decisions to benefit themselves and not the poor majority (Barger, 2016).

1.2.2 The Sustainable Development Goals

Another dimension of inclusivity in cities revolves around the concept of sustainability. This is a powerful concept that revolves around achieving and supporting a process over time without depletion. When applied to developmental circles, the emphasis is on ecological equilibrium that maximises efficiency. It points to co-existence between the natural environment and humanity with the intention of continuity in future. This comes from the realisation that natural resources, when depleted, are not easily replenishable. In the urban context, the World Bank (2019) defines sustainable cities as those that are able to adapt to, mitigate and promote economic, social and environmental change. The focus is on change of human behaviour in terms of the use of current resources without putting unnecessary strain on them, while keeping the future generation in mind so that they too, can have an opportunity to benefit from them. The four key words which form the basis of the definition – *Environment, Development, Future and Human Behaviour* – all converge in a complex way that brings diversity to the meaning of sustainable cities.

A comprehensive position about sustainable development in cities is spelt out by the United Nations' Sustainable Development Goals (SDGs) of 2015, built on the backdrop of the United Nations Development Goals (DGs) of 2000. This became the platform upon which 193 countries agreed to achieve prosperity by protecting the planet through strategies that build economic growth and address a wide range of social needs (Short, 2015). The agenda to advance sustainability revolves around three Es – *Economy, Ecology and Equity*. In essence, the need to achieve economic growth should not be at the detriment of the environment, and the benefits so achieved should be enjoyed by everyone. This explains why the New Urban Agenda (UN-Habitat) extends the arguments on the right to the city by bringing on board the issue of integrating equity to the developmental agenda. This revolves around social justice in order to ensure access to the public sphere and extend opportunities to everyone. Hence, another definition of sustainable cities (Sustainability for All, 2019) is:

> *living spaces built to protect the environment, defend social justice and promote inclusive development which does not leave anyone behind.*

What is more intriguing is that out of the 17 SDGs, 11 of them provide a shared vision on cities which, in one way or the other, affect urban residents, including the vulnerable, as shown in Table 1.1.

An overview of these SDGs shows the complexity of building sustainable cities both as functional physical constructs and as habitable human environments.

1.2.3 The Anchors of Inclusive Cities

The right to the city, supported by sustainable development goals, should ultimately provide a suitable human environment that is responsive to socio-economic needs. The discourse on inclusive cities is one of those areas that are increasingly generating enormous debate in academic circles, communities and among policy makers. Emerging narratives around inclusive cities point to diverse definitions and areas of interest such as conceptualisations, mathematical models, cultural diversity, spatial accessibility, economic growth, public policy and sustainable development, among others (Short, 2021). A wide range of empirical evidence on the built environment, specifically on cities, is available to support this diversity, as shown in our previous publication on *Inclusive Cities in Southern Africa* (Magidimisha-Chipungu and Chipungu, 2021). Therefore, this observation on diversity of inclusive cities is not misplaced but is a pointer to the intricacy of issues that are built into the urban environment and which affect both human domains and the physical environment.

For that reason, there are competing definitions of what comprise inclusive cities, as summarised in Table 1.2.

Table 1.1 Sustainable Development Goals and the Urban Agenda

SDG No.	Key Theme	Relevance to the Urban Agenda
1	No poverty	• Ensure that all have equal rights to economic opportunities and access to basic services. • Build the resilience of the poor, reduce their exposure and vulnerability to climate related shocks and disasters.
2	Zero hunger	• Increase investment in rural infrastructure
5	Gender equality	• Eliminate all forms of violence against women and girls in public and private places.
6	Clean water and sanitation	• Achieve universal and equitable access to safe and affordable drinking water for all. • Achieve access to adequate and equitable sanitation and hygiene for all.
7	Affordable and clean energy	• Double the global rate of improvement in efficiency.
8	Decent work and economic growth	• Promote policies that support productive activities, decent job creation, entrepreneurship, creativity and innovation; and encourage the formation and growth of SMEs. • Achieve full and productive employment and decent work for all.
9	Industry, innovation and infrastructure	• Develop reliable, sustainable infrastructure to support economic development and human well-being.
12	Responsible production and consumption	• Substantially reduce waste generation.
13	Climate action	• Strengthen resilience and adaptive capacity to climate related hazards.
16	Peace, justice and strong institutions	• Significantly reduce all forms of violence and related deaths rates everywhere.

Source: United Nations (2015).

Although the definitions in Table 1.2 are not exhaustive of all definitions of inclusive cities, they provide an insight into the derivatives of what comprises inclusivity in urban space. The different facets of inclusivity can be drawn from the SDGs discussed in the preceding paragraph and in Table 1.2. What is important in this regard is to draw points of convergence which together provide opportunities and environments for better living conditions for all. This resonates with the Asian Development Bank (2014) motto that '*Liveable Cities are Inclusive Cities*' and must be socially, economically and environmentally sustainable in all urban operations. The ADB's web of inclusivity is summarised in Box 1.1.

Taken from the above discussion, it is clear that there are many attributes which contribute to the development of inclusive cities. But what further complicates the puzzle is the ability to pursue these different dimensions of inclusivity without contradictory results which might reinforce exclusion. Short (2021) takes this analysis further by mapping the different definitions

Table 1.2 The Many Definitions of Inclusive Cities

Author	Definition
Anttiroiko and de Jong (2020)	An Inclusive city is a democratically governed citizen-centric urban community where individuals and groups contribute towards urban development beyond conventional financial capital but also include, human, social, physical and natural capital which are all valuable even when they cannot be measured satisfactorily.
Kundu, Sietchiping and Kinyanjui (2020)	An inclusive city should take into account spatial, economic, social, cultural and environmental dimensions of the cities based on participatory process, push factors (subsidies and institutional framework) and societal mobilisation.
UCLG (2019)	An inclusive and accessible city is a place where everyone, regardless of their economic means, gender, ethnicity, disability, age, sexual identity, migration status or religion, is enabled and empowered to fully participate in the social, economic, cultural and political opportunities that cities have to offer.
Asian Development Bank (2014)	An inclusive city creates a safe liveable environment with affordable and equitable access to urban services, social services, and livelihood opportunities for all the city residents and other city users to promote optimal development of its human capital and ensure the respect of human dignity and equality.
United Nations Habitat (2002:5)	An inclusive city is a city that promotes growth with equity. It is a place where everyone, regardless of their economic means, gender, race, ethnicity or religion, is enabled and empowered to fully participate in the social, economic and political opportunities that cities have to offer.

Source: As specified in the table.

Box 1.1 Dimensions of Inclusivity in Cities

- Joint Strategic visions of all stakeholders through participatory planning and decision making.
- Knowledge and information sharing.
- Public participation and contribution.
- Adequate standard of living for all through cross-subsidies, social protection and gender balance.
- Sustainable use of resources.
- Business environment that attracts capital investment for all to allow economic activities.
- Resilience to global environmental and socio-economic shocks and threats.

Source: ADB (2017:4)

Table 1.3 Anchors of an Inclusive City

Anchor	Dimensions
Spatial inclusion	Equal access to public housing, transportation and public infrastructure
Social inclusion	Sustainable migration and smart participation and citizenship
Economic inclusion	Community and finance and segregation and economic regeneration
Environmental inclusion	Production and consumption in such a manner that the needs and interests of future generations are not compromised.
Political inclusion	Equity in terms of political rights and obligations before the law by citizens in their state.

Source: Author (2022 – Adapted by Author from Short, 2021).

and interpretations of inclusive cities with the intention of creating key anchors that reflect the main clusters. These anchors stem from the understanding that the concept of inclusive cities is urban-development oriented with a focus on eliminating urban exclusion, inequality and discrimination – all this is driven by various mechanisms such as technological innovation, participation of local stakeholders, vital resources and tolerance for those who are willing to contribute towards inclusive urban prosperity. This in turn results in a complex web of multiple spatial, social, political, environmental and economic factors. While these appear to be distinct from each other, there are mutual synergies which make them applicable to the urban environment in a holistic manner (Short, 2021). Table 1.3 provides the anchors of an inclusive city with its dimensions.

In summary, it can be argued that building inclusive cities is a complicated process which requires the participation of all stakeholders irrespective of their position or status in society. However, where do immigrants fit in this discourse of inclusivity? The literature on inclusive cities and the principles that underpin it are very clear – the emphasis is on non-discrimination irrespective of status. The literature revolves around the principle of '*Leave No One Behind*'. This is the core foundation of an inclusive city which requires urban authorities and the state at large to build their policies with everyone in mind regardless of their status. But it does not necessarily imply that governments should disregard their sovereignty in how they respond to immigrants. The emphasis is on avoiding discriminatory policies in relation to accessing opportunities and services available in cities. Hence, Anttiroiko and de Jong (2020) conclude that an inclusive city can only thrive when obstacles to dignity, freedom, self-expression and value creation have been removed. This, in addition, should be understood in the context of vulnerability where most of the urban poor (including those living with disabilities, the elderly, women and children) find themselves trapped in. Urban immigrants (especially those who are not documented) fall in this same category because they are always marginalised. But then, who are urban immigrants? The response to this question is found in Section 1.3 which deliberates on the notion of immigrants.

1.3 The Notion of Immigrants

At the centre of the debacle to demystify inclusivity in cities are urban immigrants who are perceived as outsiders. This perception of immigrants as outsiders has not only robbed them of their rights to the city but also marginalised them to such an extent that they are part of the vulnerable groups in urban societies. But who are the rightful citizens of cities? More so, who are immigrants? In the context of this book, the focus is on people who belong to countries other than where they are staying.

1.3.1 Who Are Immigrants in Cities?

The term 'immigrants' (or migrants) is borrowed from the Latin word *migrare* which simply means wanderer. It first emerged in modern literature around the 17th century with reference to the movement of people between nations. The UN-Habitat (2018) provides an elaborate definition of migrants by noting that the term migrant applies to:

> any person who is moving or has moved across an international border or within a state away from his/her habitual place of residence, regardless of (1) the person's legal status; (2) whether the movement is voluntary or involuntary; (3) what the causes for the movement are; or (4) what the length of the stay is.

This is an umbrella term which is not covered under any international law but serves its purpose to cover the identity of people caught up in such movements. What is interesting about this definition is the acceptance of immigrants within borders of state. In this regard, urban areas emerge as centres of immigrants within and outside national states. Dwelling on the aspect of movements within national borders, it is interesting to note that approximately 740 million people have moved and settled into cities (UN 2014). However, the focus in this book goes beyond internal immigrants and shifts to those people who cross national borders into other countries. Gimeno-Feliu et al. (2019), provides further clarification on the definition of 'international immigrant' from two conspicuous positions:

- *An immigrant as foreign-born*: This refers to a person who was born in a country other than their current country of residence. However, this definition does not include second and third generations of immigrants since they are considered to be independent of their parents or grant parents' migratory history.
- *An immigrant as a foreigner or non-national*: This is a person whose national status or citizenship is of another country. This definition largely depends on the legal status of each country since this status can change.

Elaborating further on the definition of migrants, the UN-Habitat (2018) provides a typology of migrants, as shown below in Table 1.4.

Table 1.4 Typology of Immigrants

Type of Migrant	Brief Explanation
Labour migrants	People who moved from their country of origin to another one or within their own country in search of employment.
Refugees	People who flee from their country to another due to internal conflict, foreign aggression, occupation, violence, fear and/or other disturbing events that threaten their lives and/or interrupt public order.
Asylum seekers	People who flee to a country other than theirs and apply for refugee status under relevant international and national instruments and are still waiting for the decision of their application.
Internally displaced	People who are forced to flee or leave their place of residence within their country due to violence, conflicts, human rights violations, natural or human-made disasters and development projects.
Crisis-displaced people	These are international migrants who are affected by conflict and human-made disasters in a country in which they work and reside and who are displaced within the country or forced to flee to a third country or return to their own country (returnees).
Climate migrants	These are people who move to urban areas, internally or internationally, as a way to cope with the intensification of the effects of climate change and environmental degradation and the decline in agriculture production in order to diversify their income or find employment opportunities that are not reliant on the environment.
Gentrified or expelled migrants	People displaced from their land, home or habitual place of residence by land grabbing deals, large infrastructure projects, urban renewal programmes and or market forces and powerful groups and who do not fit under the traditional categories of migrants, refugees or IDPs
Other migrants	These are, for instance, students and families of labour migrants.

Source: Author 2022 (Adapted from UN-Habitat 2018a:7).

1.3.2 *Immigrants and the City*

It will be an illusion to think that migration is not an urban affair. People are migrating to cities every day and this is happening within and between countries. Existing statistics show that in the Asian and Pacific Regions, approximately 120,000 people are migrating to cities on daily basis. It is further projected that by 2050, the population of people in the same region will have increased by 63% (UN ESCAP, 2014).

Southern African cities are among the cities (such as those in East Asia and Brazil) which are experiencing higher economic growth and, therefore, also becoming key magnets for immigrants. At a global scale, it is estimated that there are 232 million international migrants (UN DESA, 2013) and 740 million internal migrants (UNDP, 2009). What is interesting about these statistics is that 50% of these international migrants are found in highly urbanised countries, where they concentrate in cities. Further projections by UN DESA (2014) indicate that approximately 2.5 billion people are expected to be in urban areas in the coming decades. This phenomenal growth is expected mainly in low- and middle-income countries of Asia and Africa.

While migrants are attracted to cities, their migration is also governed by the type of the city and the intention of their movement. High-ranking cities (known as *global cities* or *world cities*) are the major attraction of migrants. These cities are regarded as international nodes which house global economic systems through connections with international financial markets and multi-national corporations (Caglar, 2015). In terms of migration, they are characterised by a highly mobile work-force and 19% of the world's foreign-born population lives here. Among such cities are London, New York, Tokyo and Hong Kong. Global cities are followed closely by *transit cities* as centres of attraction for migrants. Transit cities are basically 'stop over' centres where migrants are forced to reside in transit to industrialised economies of the Global North or other similar regional economic hubs. They are common in countries like Morocco and Mexico. These are mainly border towns where migrants are forced to reside due to restrictive border controls and migratory policies – hence they end up being long-stay destinations. Within the Southern African region, some cities in Malawi, Zimbabwe and Mozambique are equally seen as transit cities as migrants normally target South Africa as their final destination. Lastly, migrants are also attracted to secondary cities, especially in the Global North. According to Neli et al. (2013), these are mainly smaller cities (with populations between 500,000 and 3 million) and some are not even known beyond their national or regional borders. The Demographia (2015) further argues that some migrants even move to cities lower than secondary cities because of employment opportunities, affordable housing, personal safety, medical facilities and educational facilities.

Generally, cities are socio-economic and political magnets of human settlements. For this reason, it is inevitable that people will always migrate to cities in search of opportunities. Movement to cities is an escape from the rural country-side where not only is life hard, but poverty is prevalent. Hence, the presence of opportunities in the form of jobs, health and educational facilities in cities attracts immigrants especially to those economies which perform well. From a global and regional perspective, such movements are driven by push factors in countries experiencing slow and uneven economic growth, political problems, as well as environmental and climatic instability (Skeldon, 2013). On the other hand, resourced cities with various economic opportunities provide instant survival opportunities, both economically and socially to which immigrants are attracted. It is these pull factors that are key determinants for immigrants.

The general prevailing attitude about immigrants is one of "being dependents" whose impact on the host country or city is negative. Contrary to this perception, it is not all immigrants in towns and cities who cannot survive on their own. While there are those who by nature of their status (as refugees) are highly dependent on the host government, there is a lot of evidence to equally suggest that immigrants contribute economically both to the host country and to their country of origin. Among many avenues through which

Table 1.5 Economic Benefits of Immigrants

Economic Activity	Brief Explanation
Workers	They are part of the local labour market (as both professionals and general labourers) and take jobs which locals at times are not willing to undertake. They alter cities' income distribution and influence local investment priorities. They have skills which range from unskilled labour to highly skilled workers, and have become a key driver for matching the skill demand and supply ratios around the globe.
Entrepreneurs	They create job opportunities and promote innovative change. They even employ local citizens in various sectors of their trade.
Students	They contribute towards increasing human capital and knowledge diffusion.
Consumers	They affect prices and production levels of host cities. In addition, they affect demand and supply of both domestic and foreign goods and services.
Savers	They are part of the host country's banking systems through which they send their remittances and foster their investments.
Tax payers	Depending on their economic status or jobs, they contribute to public funds through taxation.

Source: Author (Author, 2022 – Adapted from OECD/ILO (2018:19).

immigrants actively contribute to the economy of the host countries are consumption, tax payment and entrepreneurship which (directly or indirectly) contribute fiscally to the city and the country as a whole, as shown in Table 1.5. More so, it should be noted that the propensity of immigrants to contribute to the economies of both the host country and their country of origin depends on their immigration status, which in turn defines their financial, intellectual, political, cultural and social capital. The overall positive impact of immigrant contribution to the city should be evaluated against the gains of urbanisation which, among others, result in diverse policy initiations, infrastructure development and diverse economic activities. This observation is substantiated by the UN DESA (2014:29) which envisages immigrants as being resources and therefore as partners in building resilience and as agents of local development and city making. The WEF (2017) observes that in 2015 migrants contributed between US$6.4 trillion and US$ 6.9 trillion which represents 9.4% of the world's gross domestic product. It further envisages that the second generation of adult immigrants are among the population's strongest economic and fiscal contributors.

However, despite the benefits discussed in the preceding paragraph, immigration comes with its own uncertainties and challenges. The World Economic Forum (2017) argues that the first generation of immigrants are costlier to the economy of the host city than the native-born population. The impact of immigrants on host cities comes in different ways and this can be discussed from a perspective of infrastructure and social implications on society and security.

(a) **Infrastructure**

Immigration to cities negatively impacts its physical and social infrastructure since most municipalities do not plan with immigrants in mind. The situation is aggravated by the presence of undocumented immigrants who rely on the support of the host government directly and indirectly through clandestine means.

- *Housing:* In most global cities, exorbitant housing prices squeeze immigrants out of the housing market, which in turn piles pressure for government interventions. On the other hand, in most developing countries, cities are under severe constraints when it comes to housing. The presence of immigrants does not only increase housing shortage but further pushes people into informal settlements where services are miserable. Equally affected are environments where immigrants choose to live in informal settlements where environmental hazards such as deforestation, indiscriminate discarding of solid waste and soil erosion are common. Most houses are built in either flood plains or hilly environments which contribute to flooding. It is estimated that over 881 million people reside in slums, and this figure is increasing annually as immigrants continue to flock to cities in the face of failures by governments to provide adequate housing.
- *Water and sanitation:* The presence of immigrants exacerbates challenges associated with water provision as well as sanitation and solid waste collection. This is due to the fact that most local authorities do not have adequate capacity to provide these services. Migrants are among a quarter of the world population living in slums which are characterised by poverty, over-crowdedness, lack of portable water and sanitation. The lack of reticulated water supply in some informal settlements forces immigrants to obtain water from rivers and shallow wells where the water is heavily contaminated. The situation at times is worsened by illegal water connections and leakage, leading to unplanned high-water consumption rates and loss. For instance, in Mexico City, 25% of the city's water supply is lost through leakage (World Economic Forum 2017).
- *Transportation and electricity:* Like any other type of infrastructure, transportation systems and power-generating utilities equally come under strain as the demand increases due to population pressure. Similarly, the consumption of electricity (some done through illegal connections in informal settlements) is one of the problems city authorities experience. As already noted, the increase in population puts pressure on the capacity of existing networks which fail to meet the demand. Illegal connections are also known to be a risk and contribute to some of the fire hazards that frequently erupt in informal settlements.
- *Education:* Immigrants contribute towards increasing enrolment figures in schools. Hence, as the number of learners increase, so does the need for more classrooms. More so, the increase in the number of immigrants affects the availability of places at school. However, poor housing

conditions, language and cultural barriers affect the competency of migrant children in schools, resulting in some of them dropping out.
- *Health:* Most immigrants, especially in the low-income bracket, do not have adequate resources to join health insurance schemes. They rely on subsidised facilities in the host countries – a factor which stretches the host countries' health budgets and infrastructure. This is a major problem especially among undocumented immigrants in transit towns who are '*invisible*' in the sense that they are not known by local authorities, do not have permanent physical addresses of residence and are therefore not included in the national or local budgets for provision of services. The poor sanitation discussed above increases physical, mental and social health risks that worsen their rate of morbidity – a factor which further puts strain on the health system of the city.

(b) **Social and Security Issues**

Apart from issues of infrastructure, immigrants also face marginalisation and safety issues in receiving countries. In most cases, immigrants are seen as outcasts, resulting in them being excluded from certain services. On the other hand, they are labelled as criminals and their presence is linked to various societal ills.

- *Social cohesion and community integration:* At the centre of attrition between immigrants and host communities in cities is lack of social cohesion and integration. This arises out of a number of issues such as cultural barriers and competition over resources and opportunities. These problems eventually lead to hatred which results in discrimination and xenophobia. In the context of South Africa, complaints of immigrants (especially Zimbabweans and Nigerians) taking South African women and jobs had led to fierce xenophobic attacks in the past (Mphambukeli and Nel, 2018). The whole process of marginalisation results in disempowerment and social exclusion with a '*binary distinction or separation between "us and them" ... a situation which enables citizens to remain unconnected*' (Mphambukeli and Nel, 2018:95) to immigrants.
- *Safety and security:* Immigrants are always accused of breaching laws, which in turn compromises the safety and security of the host countries. Among such activities which impact on state security and safety is continual disregard for host countries' immigration policies, migrant smuggling, human trafficking through use of undesignated entry points and corrupt means to reach their intended destinations. Similarly, illegal activities in host countries such as prostitution, use of drugs and engaging in illegal mining are among the pertinent issues which receiving countries complain about. Leshoro (2022) observes that *Zama Zama* (illegal miners) mining syndicates in Johannesburg undermine police authority, and they engage in activities such as rape, illegal possession of arms and robberies.

In summary, it can be argued that urbanisation problems that affect cities are exacerbated by the presence of immigrants. However, this is inevitable given that the age of technology has revolutionised the flow of information, making it easy for immigrants to move to their destination in relatively short time. Migrants are becoming smarter in their movement as they are highly informed about their destinations. Therefore, migration, as a natural phenomenon, is here to stay – what is required is to explore means through which to minimise the negative impacts of such movements in order to make them beneficial to both immigrants and host cities.

1.4 Book Outline

This book is a follow-up on our first publication in 2018, where we discussed migration in post-colonial Southern Africa from the perspective of '*Crisis and Identity*' among migrants in the region. The discourse in this book takes the narrative further by focusing on selected cities of Southern Africa with the intention of building a holistic picture about immigration in receiving countries. The point of departure in this regard is the refocusing of the whole discussion on inclusive cities where the intention, as already pointed out, is to establish the extent to which migrants are assimilated into host cities. This stems from the perspective of cities being major attractions to migrants. As can be recalled, Southern Africa is no stranger to these movements. Traditionally, movement patterns within the region focused more on mines and farming areas with less migrants targeting cities. The need to create European towns that are exclusive and ultimate bastions for European minority settlers prevented most African migrants from moving to cities. Using various strategies such as legislation and racial discrimination, colonial masters and apartheid architects ensured that there were minimum inflows of black population into cities. The demise of the last bastion of white supremacy, the apartheid system, paved way to a free Southern Africa which has witnessed the movement of people within the region and from beyond. Existing statistics show that by mid-2020, the region was home to a population of 363.2 million and 6.4 million international migrants (Nhengu, 2022). Most immigrants are attracted to economic hubs of the region such as South Africa, Angola and Zambia where vibrant industrial developments, mining and oil wealth lure both skilled and unskilled immigrants. Unfortunately, some countries bear the brunt of these movements with South Africa being estimated to be housing approximately 2.9 million (UN DESA, 2020) immigrants. The situation has been worsened by the instability in regions such as Zimbabwe and Mozambique which generate more migrants.

It is therefore important to diagnose the state of preparedness in these Southern African cities given this surge in immigrants. Hence, the increase of migrants poses a number of questions:

- How are migrants received in these cities?
- How are these immigrants integrated into society?

- What is the reaction of the host population to immigrants?
- How are municipalities coping given the influx of these immigrants?

These and other questions form the mainstay of this book. The discourse in this book is spread over 12 chapters with each chapter dedicated to a particular theme.

The preamble into inclusivity and immigrants is provided in *Chapter 1* where the *agenda for inclusivity with immigrants in mind* is set. This is a comprehensive chapter which sets the stage by exploring the diverse meanings of the key concepts that underpin this book – inclusivity and immigrants. These are given a thorough insight since they form the basis upon which the whole book is driven. Migration patterns are changing – and they are changing very fast as immigrants target cities. The focus is not only on global cities but also on secondary and transit cities. Hence, the city emerges as the main level of analysing immigration.

Compiling this book required diverse sources of data. *Chapter 2*, 'Research Methodology', discusses the various sources of data which were used to compile this book. These included both primary and secondary data sources with both qualitative and quantitative data being incorporated. The use of secondary data sources enabled access to government publications which are crucial when it comes to interrogating the issue of immigrants. But more so, secondary data enabled the authors to engage content analysis given the nature of the topical issues that were being interrogated. Primary data, though used sparingly throughout the case studies, found more relevance in Chapter 11 in an in-depth discussion on the level of inclusiveness on South African cities and immigrants.

An insight into immigrants and cities is not conclusive without tapping into the institutional frameworks that govern people and states. 'Immigrants and the City', which is *Chapter 3* of this book, provides a platform upon which the contested issue of migrants can be analysed. It takes its clue from Chapter 1 but with emphasis on migration and with the international economic system in mind given that migration is tied to international capital. But more so, migration is responsive to institutional frameworks which operate at both national and international levels. Hence, various governments' responses in trying to assimilate migrants into their societies are responses to both local institutional requirements and international laws. This in turn affects levels of assimilation, as shown under the four models of acculturation.

The discourse on inclusive cities and immigrants discussed in Chapter 1 should also be understood in the context of regional dynamics. *Chapter 4* on the 'Southern African Region in a Historical Perspective' is a comprehensive chapter which steps into the history of the region as we try to re-trace migration patterns. The region is no stranger to the dynamics of migration as it goes back deep into the then traditional kingdoms which have since scattered around Southern Africa. The recent visit to the Zulu King's coronation (20/08/2022) by African delegates representing traditional leaderships from

Zambia, Malawi and Mozambique all point to a unique history where people were divided by colonial administrative borders. But more so, the diversity of regional movements was complicated by the search for land and mineral wealth in a bid to create industrialised cities as we see them today. The need for both skilled and unskilled labour generated movements from within the region and from as far as India and Europe. Hence, the immigration patterns we see are a *continuous lyric* of the past but now being perfected in line with contemporary dynamics.

Empirical evidence of prevailing immigration policies and laws is provided in the form of case studies from Chapter 5 to 10. These four chapters are at the core of this book as they interrogate the issue of immigrants in four Southern African capital cities. Gaborone, which is known as 'The Gem of Africa', due to the prevalence of diamond in the country, is the capital city of Botswana. The unpacking of migrant laws in Gaborone is done in *Chapter 5*. This is followed by 'a Retreat into Lilongwe', in *Chapter 6*. Lilongwe, which is the capital city of Malawi, is also the major city in the country and is called the 'City Centre'. A discussion on 'the Lush Capital of Mozambique', Maputo is offered in *Chapter 7*. This is the only city included in this book whose history and development is associated with the Portuguese influence. Its coastal location positioned the city and the country at large close to international links. This is closely followed by 'eGoli', Johannesburg (in *Chapter 8*), which historically was associated with gold mining – hence the name, which simply means gold. The discussion on *Lusaka*, the capital city of Zambia is done in *Chapter 9*. The city evolved as an administrative centre of the country buoyed by the vast copper mines of the Copperbelt Province. The last case study is on Harare in *Chapter 10*, the capital city of Zimbabwe. This symbolically is a retreat into 'the Shadows of the Sunshine City'. The city, in its historical context, was known for its splendour and beauty which mesmerised the minority settlers – hence its choice as the centre of administration by the Federation Regime discussed in Chapter 4.

The inclusion of the 'aside' on the 'Inclusion of Foreign Migrants in South Africa' in *Chapter 11* is not a mistake but a well-calculated move which was meant to provide a deep insight into reality. Extracted from in-depth interviews with foreign nationals in Durban, Cape Town and Pretoria, this chapter takes an exceptional focus to detail out the daily experiences of immigrants. It is a completely different narrative which is driven precisely by daily experiences in urban space supported by official positions mostly drawn from key informants. Above all, it speaks to the different levels of inclusivity experienced at both individual and group levels.

Reaching a consensus on diverse issues emanating from different countries and on a topical subject such as migration is a challenging undertaking. *Chapter 12* provides a platform for 'Reflections on Immigration Policies in Southern African Cities'. This is a stage where 'loose ends are tied together' with the intention of coming up with a holistic picture about migration policies. The focus in this chapter is on how immigrants are assimilated into host societies. However, this is not an easy policy intervention measure in any country given the need to balance the host population's expectations and that of immigrants. The biggest challenge which Southern African cities face is that most immigrants are in the low-income bracket and are not documented. Hence, this chapter discusses the levels of inclusivity in these cities.

This book ends with *Chapter 13* – titled the 'Epilogue'. While the intention of this chapter is to provide a summary of the major issues covered in this book, it goes beyond by giving recommendations meant to build inclusive cities. This chapter is built on the premise that migration as a natural phenomenon will always be there – hence governments need to put in place measures that will help to stabilise migrants within their cities. The recommendations provided are not cast in concrete since countries and cities have diverse socio-economic dynamics – hence all recommendations should be governed by conditions prevailing in their context.

1.5 Summary

At the heart of this book is an effort to contribute towards building cities that are inclusive and cities that respond positively to immigrants. On the one hand, it should be borne in mind that inclusive cities are fragmented and their notion of coherence as depicted in literature is difficult to find in practice. However, the reality of the matter is that inclusivity exists under different themes in different sectors. While the goal of building such inclusivity at a national level is almost impossible, the need is to look into individual policies that speak to different aspects of the city. It is through this approach that certain policy actions can be synthesised in a bid to build consistency that can be replicated and supported in other sectors. Southern Africa is diverse and the notion of migrants will continue pestering governments. The situation is aggravated by spatial differences in resources, which in turn force migrants to flock to those cities that are perceived to be performing well economically and therefore provide opportunities. It is for this reason that concrete measures to achieve inclusivity are essential if tomorrow's cities are to embrace human diversity in their growth.

Bibliography

Aalbers, M. B. & Gibb, K. 2014. *Housing and the Right to the City: Introduction to the Special Issue.* Taylor & Francis.

Anttiroiko, A.-V. & De Jong, M. 2020. *The Inclusive City: The Theory and Practice of Creating Shared Urban Prosperity.* Springer.

Asian Development Bank. 2014. *Urban Sector Group.* Cities Working Group.

Asian Development Bank. 2017. Enabling Inclusive Cities: Tool Kit for Inclusive Urban Development. https://www.adb.org/sites/default/files/institutional-document/223096/enabling-inclusive-cities.pdf

Barger, K. 2016. *Densification as a Tool for Sustainable Housing Development: A Case Study of Umhlanga High Income Area in Durban.* University of KwaZulu Natal.

Caglar, A. 2015. Urban Migration Trends, Challenges and Opportunities in Europe. *World Migration Report.*

Demographia. (2015). *Demographia World Urban Areas (Built Up Urban Areas or World Agglomerations), 11th Edition.* https://www.urbangateway.org/es/system/files/documents/urbangateway/db-worldua.pdf

Duke, J. 2010. Exploring Homeowner Opposition to Public Housing Developments. *Journal of Sociology & Social Welfare*, 37, 49.

Escap, U. 2014. *Climate Change and Migration Issues in the Pacific.* ESCAP.

Gimeno-Feliu, L. A., Calderón-Larrañaga, A., Díaz, E., Laguna-Berna, C., Poblador-Plou, B., Coscollar-Santaliestra, C. & Prados-Torres, A. 2019. The Definition of Immigrant Status Matters: Impact of Nationality, Country of Origin, and Length of Stay in Host Country on Mortality Estimates. *BMC Public Health*, 19, 1–8.

Harvey, D. 2012. *Rebel Cities: From the Right to the City to the Urban Revolution.* Verso Books.

Kundu, D., Sietchiping, R., & Kinyanjui, M. 2020. *Developing National Urban Policies.* Springer.

Lefebvre, H. 1996. *Writings on Cities*, Ed. Eleonore Kofman & Elizabeth Lebas. Blackwell.

Lefebvre, H. 2003. *The Urban Revolution.* University of Minnesota Press.

Leshoro, D. 2022. *Government Too Incompetent to Deal with Zama Zama Cancer.* City Press.

Magidimisha-Chipungu, H. H. & Chipungu, L. 2021. *Urban Inclusivity in Southern Africa.* Springer.

Mphambukeli, T. N. & Nel, V. 2018. Migration, Marginalisation and Oppression in Mangaung, South Africa. In H. Magidimisha, N. Khalema, L. Chipungu, T. Chirimambowa, & T. Chimedza (Eds.), *Crisis, Identity and Migration in Post-Colonial Southern Africa.* Springer.

Neli, E., Pugliese, A., & Ray, J. 2013. *The Demographics of Global Internal Migration.* International Organization for Migration.

Nhengu, D. 2022. Covid-19 and Female Migrants: Policy Challenges and Multiple Vulnerabilities. *Comparative Migration Studies*, 10, 1–16.

OECD/ILO. 2018. *How Immigrants Contribute to Developing Countries' Economies.* OECD Publishing.

Short, J. R. 2015. *Why Cities are a Rare Good News Story in Climate Change.* UMBC Faculty Collection.

Short, J. R. 2021. Social Inclusion in Cities. *Frontiers in Sustainable Cities*, 3, 22.

Skeldon, R. 2013. *Global Migration: Demographic Aspects and Its Relevance for Development.* United Nations.

Soja, E. 2010. *Globalization and Community: Seeking Spatial Justice.* University of Minnesota Press. http://www.ebrary.com

Sustainability For All. 2019. https://www.activesustainability.com/Constructionand-Urban-Development/Cities-Communities-Sustainable/?_Adin=02021864894

UCLG. (2019). *Inclusive and Accessible Cities in CitiesAreListening Durban United Cities and Local Governments.* UCLG.

UN DESA. 2013. International Migration 2013 Wallchart. www.un.org/en/development/desa/population/migration/publications/wallchart/docs/wallchart2013.pdf

UN Department of Economic and Social Affairs Population Division [UN DESA]. 2014. *World Urbanization Prospects: 2014 Revision.* Working Paper no. ST/ESA/SER.A/366. UN Bureau of Economic and Social Affairs.

UN-Habitat. 2018a. Migration and Inclusive Cities. A Guide for Arab City Leaders. *United Nations Human Settlements Programme.*

UN-Habitat. 2018b. *Tracking Progress towards Inclusive, Safe, Resilient and Sustainable Cities and Human Settlements*. SDG 11 Synthesis Report-High Level Political Forum 2018.

UN-Habitat. 2002. *The Global Campaign on Urban Governance: An Inventory*. Un-Habitat.

United Nations. 2014. *World Urbanization Prospects: 2014 Revision, Department of Economic and Social Affairs, Population Division*. New York.

United Nations (UN). 2015. *Transforming Our World: The 2030 Agenda for Sustainable Development*. UN.

United Nations Development Programme (UNDP). 2009. *Human Development Report 2009. Overcoming Barriers: Human Mobility and Development*. Palgrave.

WEF. 2017. Migration and its Impact on Cities. *World Economic Forum in Collaboration with PWC*.

World Bank. 2019. *Structural Transformation Can Turn Cities into Engines of Prosperity*, April 17. https://www.worldbank.org/en/news/feature/2019/04/17/structural-transformation-can-turn-cities-into-engines-of-prosperity

World Economic Forum. 2017. *Migration and Its Impact on Cities*. World Economic Forum.

2 Research Methodology

2.1 Introduction

Research methodology shows the path through which these researchers formulate their problem and objective and present their result from the data obtained during the study period. This chapter hence discusses the research methods that were used during the research process. It includes the research strategy, research design, research methodology, the study area, data sources such as primary data sources and secondary data, data collection methods like primary data collection methods, including workplace site observation data collection and data collection through desk review, data collection through questionnaires, data obtained from experts opinion, secondary data collection methods, methods of data analysis used such as qualitative data analysis, the reliability and validity analysis of the qualitative data, reliability of data, reliability analysis, validity, data quality management, inclusion criteria, ethical consideration and dissemination of the result and its utilisation approaches. The study used these mixed strategies because the data were obtained from all aspects of the data source during the study time. Therefore, the purpose of this methodology is to satisfy the research plan and target devised by the researcher. The question of whether or not a city is inclusive to migrants is based on an empirical analysis, making use of secondary data.

2.2 Research Design

The research design is intended to provide an appropriate framework for a study. A very significant decision in research design process is the choice to be made regarding research approach since it determines how relevant information for a study will be obtained; however, the research design process involves many interrelated decisions. This study adopted and used a qualitative research approach within the interpretive phenomenological research design to understand experiences from the participants' viewpoint (Leedy and Ormrod, 2013). The main characteristic of qualitative research is that it is mostly appropriate for small samples, while its outcomes are not measurable and quantifiable. Its basic advantage, which also constitutes its basic difference from quantitative research, is that it offers a complete description and

DOI: 10.4324/9781003184508-2

analysis of a research subject, without limiting the scope of the research and the nature of participant's responses (Collis and Hussey, 2003). In qualitative research individuals are generally selected to participate based on their experience of a phenomenon of interest (Speziale et al., 2011). The issue of inclusion of migrants could thus be best captured in a qualitative way where knowledge, feelings and even perceptions are expressed freely in a manner that avails more understanding.

Hence, this study employs a descriptive research design to agree on the issue of inclusive cities for immigrants in the selected cities in different Southern African countries. Saunders et al. (2009) and Miller (1991) say that descriptive research portrays an accurate profile of persons, events or situations. This design offers to the researchers a profile of described relevant aspects of the phenomena of interest from an individual, organisational and institution-oriented perspective. Therefore, this research design enabled the researchers to gather data from a wide range of respondents on the inclusivity of six cities to migrants in the respective six countries of Southern Africa. And this helped in analysing the response obtained on how it affects their integration and social-economic development.

The effectiveness of this qualitative research is heavily based on the skills and abilities of researchers, while the outcomes may not be perceived as reliable because they mostly came from the researcher's personal judgements and interpretations. Because it is more appropriate for small samples, it is also risky for the results of qualitative research to be perceived as reflecting the opinions of a wider population (Bell, 2005).

2.3 Sampling

Purposive sampling was used to select participants to the study. These participants were chosen on the basis that they would be able to provide adequate information on governance and experiences regarding immigrants in a particular city or country. Six cities from six countries (i.e. Lusaka in Zambia; Lilongwe in Malawi; Harare in Zimbabwe; Gaborone in Botswana; Johannesburg in South Africa and Maputo in Mozambique) in the Southern African region were purposively selected for the study. In addition, three more cities in South Africa were selectively chosen for detailed surveys, viz Durban, Pretoria and Cape Town). The selection of these cities is based purely on the researcher's intuition, but this is also worth justifying. These cities have experienced and continue to experience migration of different proportions as people enter and leave the country for various reasons. Some cities are in middle-income countries while other are in low-income countries. The selection of countries or cities from different economic development levels was designed to capture varying experiences as these countries or cities have different levels of opportunities and facilities, and also generally a comparatively different social environment.

The advantage of purposive sampling is that it allows the researcher to home in on people or events, which, there are good grounds for believing, will be critical for research. Instead of going for the typical instances, a cross-section or a balanced choice, the researchers concentrated on instances, which they assumed would display a wide variety – possibly even a focus on extreme cases – to illuminate the research question at hand. In this case it might not only have been economical but also have been informative in a way that conventional probability sampling could not have been (Ross, 2005).

Within this context, the participants were limited to key informants, individual immigrants and focus group discussions. According to this purposive sampling method, which belongs to the category of non-probability sampling techniques, sample members are selected on the basis of their knowledge, relationships and expertise regarding a research subject. In the current study, the sample members who were selected had special relationship with the phenomenon under investigation, sufficient and relevant work experience in the field of migration, active involvement in several migration management initiatives and partnerships, as well as proven background and understanding of raw data concerning the topic under discussion.

2.4 Data Collection Methods and Tools

In conducting fieldwork in these cities, a number of research tools and techniques for collecting data were used, including secondary and primary data collection methods.

2.4.1 Secondary Data Sources

Secondary sources of data were obtained from a variety of sources, which range from the public sector to the private sector. It was envisaged that the public sector (i.e. the City of Councils, the governments of each country, international bodies and institutions) would provide information pertaining to the migrant's population, their experiences, governance, legislation, policies and information on immigrants and how they form the urban fabric of each country. Verification of such information was done through private organisations (such as the Urban Development Corporation (UDCORP)) and non-governmental organisations (NGOs) (i.e. UN-Habitat, cooperatives, UNHCR, the International Organization for Migration (IOM), World Bank and faith-based organisations) from which significant information on migration and migrants' experiences was obtained. In addition, it has to be admitted that a lot of information that is documented was captured by the mass media in the form of newspaper reports and video clips – hence this was used to expand both secondary and primary data sources. The use of documented data provided immense insights into migration issues and comprise both historical and contemporary data sources on urban inclusiveness. In addition, the use of such a wide range of existing data provided greater scope and depth in understanding the inclusive cities concept. Such information, if reliable and accurate, further provided opportunities for replication of some issues.

However, it has already been noted above that information that is generated through mass media need to be treated with caution. It was left up to the researcher to judge the authenticity of such records given that some of them might have been deliberately distorted or unconsciously misrepresented – thus presenting a major limitation in the use of secondary data.

2.4.2 Content Analysis

Southern Africa as a region covers almost half of the African continent. This spatial extent in itself is not only a barrier to engage in direct data collection at both country and city levels but is also strenuous and expensive. The need to manoeuvre bureaucratic processes coupled with extended waiting times to get access to key informants did not make matters easy, but worsened the situation. Under such circumstances, content analysis emerged as one of the most powerful research strategies used by the authors to gather information for this book. Easy access to published materials made it convenient for authors to select appropriate content required for this book. The situation was even made easier by sifting through computers using the internet, which allowed access to a wide range of databases with online books, journal articles, conference proceedings, organisational databases and government policy and legislative documents which could be easily downloaded and examined carefully. The ability to download multiple documents on the same subject provided an opportunity to compare definitions and content in order to draw meaningful inferences upon which this book was finally compiled. Through content analysis, both qualitative and quantitative data sources were accessed which enabled the authors to build a holistic picture about developments in the Southern African region.

The search for meaningful content revolved around three key concepts – viz. *Inclusive Cities, Immigrants* and *Southern Africa* – with a focus on cities. These are briefly discussed below.

Inclusive Cities: In contemporary urban studies, this is an elusive concept whose meaning is complex. As a circuitous concept, it is made up of intricate and interconnected developmental urban terminologies that require deeper cognate processing and processes. This in turn required combing through related urban developmental terminology such as *sustainable development*, *smart cities* and *the right to the city*. Put together, the many meanings of the definition of inclusive cities provided in this book are derivatives extracted from expected and appropriate intervention measures that should manifest in the urban realm. Therefore, the ultimate and accepted dimensions of inclusivity were informed by renowned urban scholars (such as Henri Lefebvre and Edward Soja) underpinned by international organisation's developmental agendas (such as those provided by the World Bank, UN-Habitat and the Asian Development Bank). Beyond the mere meaning of the concept, its intricacy was also searched in line with the objective of the book, i.e. *urban inclusivity vis-à-vis immigrants*. This further widened the search with a focus on various publications that champion on how migrants are incorporated or excluded in host countries. Once more, publications from the World Bank, Asian Development Bank, UNESCO and UN-Habitat (among others) were the major sources of such literature. Therefore, sifting through various sources of literature in relation to inclusive cities and immigrants yielded a total of 30 publications.

Immigrants: The dynamics of modern-day societies generate a lot of movements due to both push and pull factors. The mere need to understand the dynamics of migration has generated a lot of interest from scholars, researchers, students, development agents and governments. Hence in the context of this research, the search was not only for the interpretation of the concept of immigration, but for intricate relationships that link the concept to urban development, inclusive development, the history and contemporary dynamics of Southern Africa. More so, it meant going through government databases in search of legislations, policies and projects in relation to immigrants across the selected Southern African countries (i.e. Zambia, Malawi, Zimbabwe, Mozambique, Botswana and South Africa) as well as specific immigration information which relate to their capital cities (i.e. Lusaka, Lilongwe, Harare, Maputo, Gaborone and Johannesburg). Equally important was the search for information that relates to regional bodies such as SADC, which is equally instrumental in shaping migration policies of the signatory members of this body. Therefore, the search for literature to inform the subject on immigrants was large and it yielded over 26 publications.

Southern Africa: The Southern African region, as already noted above, covers almost half of the Southern tip of the whole continent. The search for information sources for Southern Africa covered three major components:

- *Southern African countries:* This is the southernmost region of the African continent. The literature search for Southern Africa focused on six countries covered in this research, viz. Malawi, Zambia, Zimbabwe, Botswana, South Africa and Mozambique. The type of literature that was compiled from the search on these countries focused on historical information on migration, government publications in the form of policy documents and legislations that relate to migration and immigrants, as well as on inclusive cities.
- *Capital cities:* Information targeted was on the major cities of the selected Southern African countries – viz. Lusaka, Harare, Lilongwe, Gaborone, Maputo and Johannesburg. The search was for publications relating to immigrants and inclusivity with a focus on government publications such as policies and legislations.
- *Inclusivity:* This is a key theme throughout the book. Literature on inclusivity in relation to Southern Africa and the concerned cities was the target of this search.
- *Immigrants:* The issue of migration and immigrants is the focus of the book. Literature on this concept focused on the selected countries and their major cities.

- **Southern Africa as a region:** The book is on the Southern African region. Hence literature relating to this region on migration, inclusivity, regional bodies and history was sourced from different publications.

A search for literature on all these topical issues on Southern Africa yielded over 153 sources.

2.4.3 Primary Data Sources

The acquisition of primary data was done by using two key methods, which are questionnaire surveys and observations. While these two techniques formed the key methods through which data was collected, other methods, where appropriate, were also used. These two methods are briefly discussed below.

2.4.3.1 Surveys

Surveys were the most important tool for primary data collection in this research. Surveys extensively relied on the administration of questionnaires through interviews, and these were undertaken in selected cities. Using Mikkelsen (2005) typology of classification of interviews, this research made use of both individual and group interviews. Basically, three distinct interviews were used, as discussed below.

2.4.3.2 Individual Interviews

These were in-depth interviews undertaken by an opportunistic sample that was made up of purposefully selected respondents in order to obtain representative information. In-depth interviews are personal and unstructured interviews, whose aim is to identify participant's emotions, feelings and opinions regarding a particular research subject. These in-depth interviews took the interview guide approach adopted from Patton's interview classifications (Frauendorfer and Liemberger, 2010). An interview guide approach is less structured than that taken in a standardised open-ended interview. The topics and issues covered were specified in advance in an outline form but the interviewer varied the wording of the questions and the sequence in which the questions were tackled. As a result, the interview had much greater freedom to explore specific avenues of enquiry; logical gaps within the data has been anticipated and closed. The interview took on a more conversational feel while ensuring that all the topics of interest were explored.

Unstructured interviews offer flexibility in terms of the flow of the interview, thereby leaving room for the generation of conclusions that were not initially meant to be derived regarding a research subject. However, there is the risk that the interview may deviate from the pre-specified research aims and objectives. Selected respondents (who will be immigrants in most cases) in the field were required to answer questions on a wide range of issues among which are:

- Access to economic opportunities in cities
- Access to basic services, i.e., housing, education healthcare, etc.
- Social integration
- Relationship with authorities

This list is not exhaustive but is only indicative of the issues that were deliberated during fieldwork. Depending on the depth of data required, the questionnaires comprised both open-ended and close-ended questions.

2.4.3.3 Key Informants Interviews

These interviews were carried out with people who are highly knowledgeable with special information relating to migrants and to specific dynamics towards the inclusive city approach. Among the people who were included in this list were senior officials in the central and local government, officials in non-governmental organisations and academic professionals. Outside the public sector, community leaders who have been associated with different migrants-related projects or initiatives were also interviewed. Some of these key informants were identified in the field through the snowball approach since it was invariably difficult to identify all of them at the preliminary stage. Unlike in the individual interviews discussed above, no structured questionnaire were designed other than an interview guide that simply showed a list of topics to be discussed. The idea behind such an approach was to make the whole interview process more informal, thereby providing an opportunity to discuss a variety of issues to some depth. Issues under discussion revolved around the following topics:

- Access to economic opportunities in cities
- Access to basic services, i.e., housing, education healthcare, etc.
- Social integration
- Relationship of migrants with authorities
- Migration governance, legislation and policies

Again, it should be noted that this list is not comprehensive and therefore not exclusive. In fact, this approach allowed the researcher to probe the interviewee, thereby opening up new topics for further discussion.

2.4.3.4 Focus Group Discussions (FGDs)

FGDs were undertaken on certain topical issues under the facilitation of the researcher. The target of these FGDs was the application of the inclusive cities concept to that city's context. The purpose of incorporating FGDs in this research was to allow for the generation of a large body of information by a group of people. In addition, it provided a platform for authenticating views emanating from various sources during fieldwork. This is in line with Patton's (1990) submission that the whole idea of FGDs is not to reach a consensus on the issues being deliberated or to solve specific problems, but to provide a platform for brainstorming, thereby equipping the researcher with more insight on topical issues. Here, a broad range of issues as noted above under key informant interviews were deliberated.

2.4.3.5 Observation

Observation is an old and traditional research tool that will always withstand the change of time. Frankfort-Nachmias and Nachmias (1996) actually argue that 'social science research is rooted in observation – it begins and ends with empirical observation'. Indeed it had already been alluded to in the preamble of this section that the way occurrences unfold in cities is spontaneous and documentation of these occurrences is sometimes and somehow very subjective. In some instances, because of the highly political nature of the manner in which they happen or unfold, some issues are labelled sensitive and are therefore shrouded in secrecy. It is therefore only through observation that the reality can be unlocked and confirmed. This approach allowed developments to be observed in their natural settings without any contamination from lapse of memory or censorship. The value of this approach is that it was used in combination with other techniques such as FGDs and interviews in recording information. The authenticity of some of the information from other sources was also verified through observation.

2.5 Data Analysis

The analytical procedures followed entailed transcribing and translating interviews, going through transcripts and highlighting significant statements (horizonalisation), grouping clustered statements into themes, coding, removing overlapping and repetitive statements, establishing themes, and writing textural description on the varying experiences of migrants. The structural descriptions were then linked to make sense of the underlying meanings, experiences and linkages made with the wider literature.

In the case of information gathered from migrants in Pretoria (Tshwane, Cape Town and Durban (eThekwini)), information gathered from migrants, whether through telephone or face-to-face interviews was initially recorded on a hardcopy of the structured survey questionnaire. Subsequently, the information was captured online via a google form. The survey was conducted mostly in English and, where possible, in French or Swahili or any other applicable African languages.

Descriptive data analysis, sentimental analysis and correlation analysis were done using the software SASPy (version 4.10). All quantitative data was checked for normality using Anderson-Darling or a Shapiro-Wilk test and a

28 Research Methodology

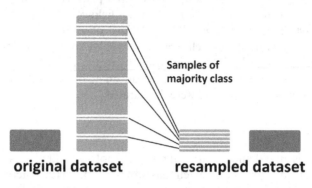

Figure 2.1 Data Analysis

normal probability plot generated. The homogeneity of variances was tested. Unbalanced analysis was applied to minimise residual errors. Synthetical observations of the larger data set (majority) class base were selected using algorithms of Modified Synthetic Majority Under-Sampling Technique as shown in Figure 2.1. The Tshwane sample was 39%, relative to Cape Town and eThekwini at 32% and 29%, respectively. The Tshwane data was the majority and the algorithms under-sampled. The Normalized Gini Score was calculated.

Pearson's (r) and the Pearson product-moment correlation coefficient (PPMCC) were generated to determine the strength of relationships among the measured variables. The correlation coefficient ranges between −1.0 and 1.0, with above zero indicating positive correlations, zero indicating no relationship and below 1.0 indicating inverse or negative relationships.

Sentimental analysis was done in SASPy to determine the ratio of negative and positive engagements within the sample by analysing bodies of text, such as comments, to obtain insights of migrants on integration. The Natural Language Toolkit (NLTK) was applied for processing and analysing text.

2.6 Trustworthiness of the Research

Issues of trustworthiness in qualitative research are always debatable. Therefore, to ensure this is addressed we undertook at least five steps (credibility, transferability, dependability, conformability and subjectivity) as advised by Lincoln et al. (2011), among others.

In terms of *credibility*, we ensured that the participants felt that the findings represented their experiences. For example, we read out the transcript to them and asked whether the views we had captured represented their views accurately or not. Where there were doubts or discontent, we rephrased the expressions to suite the participants. In addition, we tried to prolong our engagement. In some cases where clarity was needed, we returned to conduct follow-up interviews. Lastly, we allowed participants to correct the narrative, where needed, once the transcripts were read to them.

Regarding the *transferability* of the findings, being conscious that they were not generalisable to a larger population, we nevertheless endeavoured to ensure rigour by collecting rich data. We took effort to describe the participants' responses as thickly as possible. This permitted us to make informed interpretations, which could allow transferability of the findings to other contexts and thus reach naturalistic generalisation.

To ensure *dependability* we asked ourselves whether similar findings would be arrived at if someone followed our methods to conduct the same study. To answer this, we used multiple methods of data collection. We supplemented the interviews with document analysis where we looked at the migrants' experiences to corroborate what participants had told us. We found this approach very helpful (Guba and Lincoln, 1994).

To guard against our own biases, motivations, interests and perspectives (*subjectivity*) we subjected our findings to an audit and produced a more transparent report – a product of participants and our responses. However, we were not in search for objectivity. Thus, we used our subjectivity as a resource rather than a deficit in order to understand the migrants' experiences and the meaning they gave to their experiences. We engaged in what others (Holmes, 2017, Xie, 2018) call inter-subjectivity (the state between objectivity and subjectivity).

Bibliography

Aalbers, M. B. & Gibb, K. 2014. *Housing and the Right to the City: Introduction to the Special Issue*. Taylor & Francis.

Aboobaker, S. 2015. Foreign-Owned Shops in Alexandra Looted. https://www.iol.co.za/news/south-africa/gauteng/foreign-owned-shops-in-alexandra-looted-1847071#.Vt36fsgqqko

Abrahams, C. & Everatt, D. 2019. City Profile: Johannesburg, South Africa. *Environment and Urbanization ASIA*, 10, 255–270.

Abrahams, C., Everatt, D., Van Den Heever, A., Mushongera, D., Nwosu, C., Pilay, P., Scheba, A. & Turok, I. 2018. South Africa: National Urban Policies and City Profiles for Johannesburg and Cape Town. Johannesburg. http://www.centreforsustainablecities.ac.uk

Amankwaa, A. A. 1995. The World Economic System and International Migration in Less Developed Countries: An Ecological Approach. *International Migration (Geneva, Switzerland)*, 33, 93–114.

Amit, R. 2010. Lost in the Vortex: Irregularities in the Detention and Deportation of Non-Nationals in South Africa. Report for FMSP. Wits University, Johannesburg.

Anttiroiko, A.-V. & De Jong, M. 2020. *The Inclusive City: The Theory and Practice of Creating Shared Urban Prosperity*. Springer.

Arrighi, G., Aschoff, N. & Scully, B. 2010. Accumulation by Dispossession and its Limits: The Southern Africa Paradigm Revisited. *Studies in Comparative International Development*, 45, 410–438.

Ashby, N. J. 2010. Freedom and International Migration. *Southern Economic Journal*, 77, 49–62.

Asian Development Bank. 2014. *Urban Sector Group*. Cities Working Group.
Asian Development Bank. 2017. *Enabling Inclusive Cities. A Tool Kit for Urban Development*.
Awumbila, M. 2017. Drivers of Migration and Urbanization in Africa: Key Trends and Issues. *United Nations Expert Group Meeting on Sustainable Cities, Human Mobility and International Migration*. Septermer 7–8, 2017, Legon, Ghana.
Barger, K. 2016. *Densification as a Tool for Sustainable Housing Development: A Case Study of Umhlanga High Income Area in Durban*. University of KwaZulu Nata.
Beegle, K. & Poulin, M. 2013. Migration ant The Transition to Adulthood in Contemporary Malawi. *The Annals of the American Academy of Political and Social Science*, 648, 38–51.
Bell, J. 2005. *Doing Your Research Project: A Guide for First Researchers in Education, Health and Social Science*. Berkshire: Open University Press.
Berry, J. W. 1997. Immigration, Acculturation, and Adaptation. *Applied Psychology*, 46, 5–34.
Berry, J. W., Phinney, J. S., Sam, D. L. & Vedder, P. 2006. Immigrant Youth: Acculturation, Identity, and Adaptation. *Applied Psychology*, 55, 303–332.
Bertocchi, G. & Strozzi, C. 2008. International Migration and the Role of Institutions. *Public Choice*, 137, 81–102.
Brelsford, W. V. 1960. *Handbook to the Federation of Rhodesia and Nyasaland*. Cassell & Company, Ltd.
Caglar, A. 2015. Urban Migration Trends, Challenges and Opportunities in Europe. *World Migration Report*.
Callamard, A. 1994. Refugees and Local Hosts: A Study of the Trading Interactions between Mozambican Refugees and Malawian Villagers in the District of Mwanza. *Journal of Refugee Studies*, 7, 39–62.
Campbell, E. & Crush, J. 2012. *Unfriendly Neighbours: Contemporary Migration from Zimbabwe to Botswana*. The Southern African Research Centre, Queen's University, Canada and the Open Society Initiative for Southern Africa (OSISA).
CCJPZ. 2017. Zimbabwe. [Online]. https://www.peaceinsight.org/en/organisations/the-catholic-commission-for-justice-and-peace-in-zimbabwe-ccjpz [Accessed 14/08/2022].
Chereni, A. & Bongo, P. P. 2018. *Migration in Zimbabwe: A Country Profile 2010–2016*. International Organization for Migration (IOM).
Chigavazira, B., Phillime, F., Kayula-Lesa, G., Shindondola-Mote, H., Frye, I., Nhampossa, J., Kaulem, J., Ramotso, M., Kafunda, M. & Mambea, S. 2012. Access to Socio-Economic Rights for Non-Nationals in The Southern African Development Community. *Open Society Initiative for Southern Africa*, 1–97. http://spii.org.za/wp-content/uploads/2019/05/Access-to-Socio-Economic-Rights-for-Non-Nationals-in-the-Southern-African-Development-Community.pdf
Chikanda, A. & Crush, J. 2016. The Geography of Refugee Flows to Zimbabwe. *African Geographical Review*, 35, 18–34.
Chinyemba, J. 2017. *Undocumented Immigration in Zambia: A Case Study of Lusaka City*. University of South Africa.
Chipkin, C. M. 1993. *Johannesburg Style: Architecture & Society, 1880s–1960s*. David Philip Publishers.
Chipungu, L. 2018. Migrant Labour and Social Construction of Citizenship in Lesotho and Swaziland. In *Crisis, Identity and Migration in Post-Colonial Southern Africa*. Springer.

Chipungu, L. & Adebayo, A. 2013. The Policy-Planning Divide: An Evaluation of Housing Production in the Aftermath of Operation Murambatsvina in Zimbabwe. *Journal of Housing and the Built Environment*, 28, 381–396.

Chipungu, L., Kamuzhanje, J., Makonese, E. D. & Magidimisha-Chipungu, H. H. 2022. Leapfrog Developments–Gaps, Challenges and Opportunities in Urban Development: Cases of Harare and Durban. *Journal of Urban Systems and Innovations for Resilience in Zimbabwe-Jusirz*, 4, 45–65.

Chipungu, L. & Magidimisha, H. H. 2020. *Housing in the Aftermath of the Fast Track Land Reform Programme in Zimbabwe*. Routledge.

Christopher, A. 1983. From Flint to Soweto: Reflections on the Colonial Origins of the Apartheid City. *Area*, 15, 145–149.

Clubofmozambique. 2017. Hardship for Refugees from Mozambique at Zimbabwe's Tongogara Refugee Camp. https://clubofmozambique.com

Coderre-Proulx, M., Campbell, B. & Mandé, I. 2016. *International Migrant Workers in the Mining Sector*. Geneva, Switzerland: International Labour Office.

Cohen, R. 2008. *Global Diasporas: An Introduction*. Routledge.

COJ. 2011. *Integrated Development Plan (IDP)*. City of Johannesburg Press Johannesburg.

Collins, J. 1969. *Lusaka: The Myth of the Garden City*. University of Zambia, Institute for Social Research.

Collis, J. & Hussey, R. 2003. *Business Research: A Practical Guide for Undergraduate and Postgraduate Students*. Palgrave Macmillan.

Craissati, D., Banerjee, U. D., King, L., Lansdown, G. & Smith, A. 2007. *A Human Rights Based Approach to Education for All*. Unicef.

CRS. 2019. *Zimbabwe*. [Online]. https://www.vaticannews.va/en/africa/news/2020-02/caritas-zimbabwe-feeds-300-000-drought-affected-persons.html [Accessed 14/08/2022].

Crush, J., Chikanda, A. & Skinner, C. 2015. *Mean Streets: Migration, Xenophobia and Informality in South Africa*. African Books Collective.

Crush, J. & Tawodzera, G. 2011a. Medical Xenophobia: Zimbabwean Access to Health Services in South Africa. *Journal of Ethnic and Migration Studies*, 40, 655–670.

Crush, J. & Tawodzera, G. 2011b. Right To The Classroom: Educational Barriers For Zimbabweans In South Africa. In *Southern African Migration Programme: Open Society Initiative For South Africa*. Migration Policy Series No 56. Institute for Democratic Alternatives in South Africa.

Crush, J., Williams, V. & Peberdy, S. 2005. Migration in Southern Africa. In *Policy Analysis and Research Programme of the Global Commission on International Migration*. Global Commission on International Migration.

CSO. 2005. *2001 Population Census Atlas: Botswana*. Gaborone: Centrals Statistics Office.

Datta, K., McIlwaine, C., Evans, Y., Herbert, J., May, J. & Wills, J. 2006. *Work and Survival Strategies among Low-Paid Migrants in London*. London: Queen Mary, University of London.

De Klerk, N. 2015. Looters Target Foreign-Owned Shops in Joburg. *News24*.

De Sousa Santos, B. 2018. *The End of the Cognitive Empire: The Coming of Age of Epistemologies of the South*. Duke University Press.

Ditshwanelo. 2005. Rights Of Minority Groups. *Botswana Centre for Human Rights, Working Paper*. Ditshwanelo.

Dodson, B. 2010. Locating Xenophobia: Debate, Discourse, and Everyday Experience in Cape Town, South Africa. *Africa Today*, 56, 2–22.
Donnelly, J. 2004. Zimbabwe Woes Spill Across Border. *Zw News* (2 March 2004), 2.
Duke, J. 2010. Exploring Homeowner Opposition to Public Housing Developments. *Journal of Sociology & Social Welfare*, 37, 49.
Englund, H. 2001. The Politics of Multiple Identities: The Making of a Home Villagers' Association in Lilongwe, Malawi. In *Associational Life in African Cities: Popular Responses to the Urban Crisis*. The Nordic Africa Institute, 90–106.
Englund, H. & Nyamnjoh, F. B. 2004. *Rights and the Politics of Recognition in Africa*. Zed Books.
Euromonitor. 2017. *City Review: Johannesburg City Review*. Euromonitor International.
Fargues, P. 2017. International Migration and Education—A Web of Mutual Causation. In *Think Piece Prepared For The 2019 Global Education Monitoring Report Consultation*. Paris: UNESCO.
Fatti, C. E. & Vogel, C. 2011. Is Science Enough? Examining Ways of Understanding, Coping with and Adapting to Storm Risks in Johannesburg. *Water SA*, 37, 57–65.
Frankfort-Nachmias, C. & Nachmias, D. 1996. *Research Methods in the Social Sciences*. London: Edward Arnold. Community Watershed Management in Semiarid India, 187.
Frauendorfer, R. & Liemberger, R. 2010. *The Issues and Challenges of Reducing Non-Revenue Water*. Asian Development Bank.
Gann, L. H. 1964. *The Growth of a Plural Society: Social, Economic and Political Aspects of Northern Rhodesian Development 1890–1953, with Special Reference to the Problem of Race Relations*. University of Oxford.
Geis, W., Uebelmesser, S. & Werding, M. 2013. How do Migrants Choose their Destination Country? An Analysis of Institutional Determinants. *Review of International Economics*, 21, 825–840.
Ghelli, T. 2016. Potatoes, Pigs and Poultry: Changing the Game for Refugees. www.unhcr.org
Gimeno-Feliu, L. A., Calderón-Larrañaga, A., Díaz, E., Laguna-Berna, C., Poblador-Plou, B., Coscollar-Santaliestra, C. & Prados-Torres, A. 2019. The Definition of Immigrant Status Matters: Impact of Nationality, Country of Origin, and Length of Stay in Host Country on Mortality Estimates. *BMC Public Health*, 19, 1–8.
Goodwin-Gill, G. S. & McAdam, J. 2007. *The Refugee in International Law, 3*. Baskı, New York: Oxford Yayınevi.
Götz, G. 2004. The Role of Local Government towards Forced Migrants. In *Forced Migrants in the New Johannesburg: Towards a Local Government Response*. Johannesburg: Forced Migration Studies Programme.
Greenburg, J. & Polzer, T. 2008. Migrant Access to Housing in South African Cities. In *Migrants' Rights Monitoring Project*. Johannesburg: Witwatersrand University (Forced Migration Studies Programme Special Report No 2).
Guba, E. G. & Lincoln, Y. S. 1994. Competing Paradigms in Qualitative Research. *Handbook of Qualitative Research*, 2, 105.
Gushulak, B. D. & Macpherson, D. W. 2006. *Migration Medicine and Health: Principles and Practice*. PMPH-USA.
Gutmann, A. 1994. *Multiculturalism: Examining the Politics of Recognition*. Princeton University Press.
Hacker, K., Anies, M., Folb, B. L. & Zallman, L. 2015. Barriers to Health Care for Undocumented Immigrants: A Literature Review. *Risk Management and Healthcare Policy*, 8, 175.

Hall, R. 1965. *Kaunda-Founder of Zambia*. Longmans.
Haralambos, M. & Holborn, M. 1995. *Sociology. Themes and Perspective* 4th Edition. London: Collins Educational.
Harvey, D. 2012. *Rebel Cities: From the Right to the City to the Urban Revolution*. Verso Books.
Harvey, D. 2020. The Condition of Postmodernity. In *The New Social Theory Reader*. Routledge.
Hawker, D. 2015. South African Businesses Hurt by Jeppestown Looting. https://www.enca.com/south-africa/south-african-businesses-affected-jeppestown-looting
Helensuzman Foundation. 2019. South Africa.
Hirschman, C. 1982. Immigrants and Minorities: Old Questions for Mew Directions in Research. *International Migration Review*, 16, 474–490.
Hlubi, P. 2015. Foreign-Owned Shops in Snake Park and Doornkorp Were Targeted by Looters. *Enews*, 20 January 2015.
Holland, H. & Roberts, A. 2002. *From Jo'burg To Jozi: Stories About Africa's Infamous City*. Penguin Global.
Holmes, J. 2017. Reverie-Informed Research Interviewing. *The International Journal of Psychoanalysis*, 98, 709–728.
IOM. 2019. Migration in Zambia: A Country Profile 2019. https://www.zambiaimmigration.gov.zm/Wp-Content/Uploads/2021/05/Zambia-Migration-Profile-2019.Pdf
IOM. 2021. International Organisation for Migration (Iom) Zimbabwe National Country Strategy 2021–2024. [Online]. https://reliefweb.int/report/zimbabwe/iom-zimbabwe-national-country-strategy-2021-2024 [Accessed 21/02 2022].
Jacobsen, K. 2001. *The Forgotten Solution: Local Integration for Refugees in Developing Countries*. UNHCR.
Jacobsen, K. 2004. Microfinance in Protracted Refugee Situations: Lessons from the Alchemy Project. *A Paper Presented in Alchemy Workshop Held in Maputo*, Mozambique.
Kabwe-Segatti, A. W. 2008. *Migration in Post-Apartheid South Africa: Challenges and Questions to Policy-Makers*. Agence Française De Développement (Afd), Département De La Recherche.
Kainja, G. 2012. A Call for a Multi-Sector Approach Against People Smuggling/Illegal Migration: A Case Study of Malawi. *Research Paper, Research and Planning Unit, Malawi Police Service*, Lilongwe, Malawi.
Kibreab, G. 1999. Revisiting the Debate on People, Place, Identity and Displacement. *Journal of Refugee Studies*, 12, 384.
Kloukinas, P., Novelli, V., Kafodya, I., Ngoma, I., Macdonald, J. & Goda, K. 2020. A Building Classification Scheme of Housing Stock in Malawi for Earthquake Risk Assessment. *Journal of Housing and the Built Environment*, 35, 507–537.
Knox, P. & Gutsche, T. 1947. *Do You Know Johannesburg? [With Illustrations and a Street Plan.]*. Unie-Volkspers.
Kowet, D. K. 1978. *Land, Labour Migration and Politics in Southern Africa: Botswana*. Lesotho and Swaziland: Nordiska Afrikainstitutet.
Laher, H. 2008. *Antagonism toward African Immigrants in Johannesburg, South Africa: An Integrated Threat Theory (ITT) Approach*. University of the Witwatersrand Johannesburg.
Landau, L. 2011. *Contemporary Migration to South Africa: A Regional Development Issue*. World Bank Publications.

Landau, L. & Polzer, T. 2007. *Xenophobic Violence, Business Formation, and Sustainable Livelihoods: Case Studies of Olievenhoutbosch and Motherwell, Johannesburg.* University of the Witwatersrand. Forced Migration Studies, University of Witwatersrand, 1–23.

Landau, L. & Polzer, T. 2008. Working Migrants and South African Workers: Do they Benefit Each Other? *South African Labour Bulletin*, 32, 43–45.

Landau, L. B. & Jacobsen, K. 2004. Refugees in the New Johannesburg. *Forced Migration Review*, 9, 44–46.

Landau, L. B., Segatti, A. & Misago, J. P. 2011. *Governing Migration & Urbanisation in South African Municipalities: Developing Approaches to Counter Poverty and Social Fragmentation.* South Africa Local Government Association (Salga).

Leedy, P. & Ormrod, J. 2013. The Nature and Tools of Research. *Practical Research: Planning and Design*, 1, 1–26.

Lefebvre, H. 1996. *Writings on Cities*, Ed. Eleonore Kofman and Elizabeth Lebas. Oxford: Blackwell.

Lefebvre, H. 2003. *The Urban Revolution*, University of Minnesota Press.

Lefko-Everett, K. 2004. Botswana's Changing Migration Patterns. https://www.migrationpolicy.org/article/botswanas-changing-migration-patterns

Lesetedi, G. N. & Modie-Moroka, T. 2007. Reverse Xenophobia: Immigrants Attitudes towards Citizens in Botswana. In *African Migrations Workshop: Understanding Migration Dynamics in the Continent.* Ghana: Centre for Migration Studies, University of Ghana, Legon-Accra.

Leshoro, D. 2022. *Government Too Incompetent to Deal with Zama Zama Cancer.* City Press.

Lesthaeghe, R. J. 1989. *Reproduction and Social Organization in Sub-Saharan Africa.* University of California Press.

Levert, S. 2007. *Botswana*. Marshall Cavendish.

Lincoln, Y. S., Lynham, S. A. & Guba, E. G. 2011. Paradigmatic Controversies, Contradictions, and Emerging Confluences, Revisited. *The Sage Handbook Of Qualitative Research*, 4, 97–128.

Lindblom, C. 2018. The Science of 'Muddling Through'. In *Classic Readings in Urban Planning*. Routledge.

Lloyd-Jones, T. & Rakodi, C. 2014. *Urban Livelihoods: A People-Centred Approach to Reducing Poverty*. Routledge.

Lupote, I. S. 2020. *The Role of the Media in Curbing Unregistered Immigrants in Zambia*. Cavendish University.

Macpherson, D. W., Gushulak, B. D. & Macdonald, L. 2007. Health and Foreign Policy: Influences of Migration and Population Mobility. *Bulletin of the World Health Organization*, 85, 200–206.

Magidimisha, H. H. & Chipungu, L. 2019. *Spatial Planning in Service Delivery: Towards Distributive Justice in South Africa*. Springer.

Magidimisha-Chipungu, H. H. & Chipungu, L. 2021. *Urban Inclusivity in Southern Africa*. Springer.

Magubane, B. 1975. The Native Reserves. In Hi Safa, and Bm Du Toit (Eds.), *Bantustans and the Role of the Migrant Labour System in the Political Economy of South Africa*, 1, 975.

Majale, M. 2001. Towards Pro-Poor Regulatory Guidelines for Urban Upgrading. A Review of Papers Presented at the International Workshop on Regulatory Guidelines for Urban Upgrading held at Bourton-on-Dunsmore, 17–18.

Makhema, M. 2009. Social Protection for Refugees and Asylum Seekers in the Southern Africa Development Community (SADC). *The World Bank, Social Protection and Labour Discussion Paper*, 15, 3–39.

Maki, H. 2010. Comparing Developments in Water Supply, Sanitation and Environmental Health in Four South African Cities, 1840–1920. *Historia*, 55, 90–109.

Malauene, M. M. & Landau, L. 2004. *The Impact of the Congolese Forced Migrants Permanent Transit' condition on their Relations with Mozambique and its People*. University of the Witwatersrand.

Manda, M. 2013. *Malawi Situation Of Urbanisation Report*. Lilongwe, Malawi: Government of Malawi.

Mandy, N. 1984. *A City Divided: Johannesburg and Soweto*. New York: St. Martin's Press.

Mangezvo, P. L. 2018. Catechism, Commerce and Categories: Nigerian Male Migrant Traders in Harare. In *Forging African Communities*. Springer.

Maphosa, F. 2005. *The Impact of Remittances from Zimbabweans Working in South Africa on Rural Livelihoods in the Southern Districts of Zimbabwe*. Citeseer.

Martinez, R., Buntin, J. T. & Escalante, W. The Policy Dimensions of the Context of Reception for Immigrants (and Latinos) in the Midwest. Cambio De Colores (10th: 2012: Kansas City, Mo.). Cambio De Colores: Latinos in the Heartland: Migration and Shifting Human Landscapes: Proceedings of the 10th Annual Conference: Kansas City, Missouri, June 8–10, 2011. Columbia, MO: University of Missouri, 2012. Cambio Center.

Masisi, P. 2009. Overstaying Immigrants' Burden Center. *Daily News*, 17.

Matiwane, Z. 2022. Ramaphosa Says Mec's Foreign Rant Could have been Handled Differently. *Times Live*.

Mbhele, M. S. 2016. *Exploring Schooling Experiences and Challenges of Immigrant Learners in a Multilingual Primary School*. University of KwaZulu Natal.

McConnell, C. 2009. Migration and Xenophobia in South Africa. *Conflict Trends*, 2009, 34–40.

Meyer-Weitz, A., Oppong Asante, K. & Lukobeka, B. J. 2018. Healthcare Service Delivery to Refugee Children from the Democratic Republic of Congo Living in Durban, South Africa: A Caregivers' Perspective. *BMC Medicine*, 16, 1–12.

Mhlanga, J. & Zengeya, R. M. 2016. Social Work with Refugees in Zimbabwe. *African Journal of Social Work*, 6, 22–29.

Miamidian, E. & Jacobsen, K. 2004. Livelihood Interventions for Urban Refugees. Paper Written for Alchemy Project Workshop, Maputo.

Migrants/Refugees. 2021. *Migration Profile, Republic of Zambia*. Vatican City: Migrants & Refugees Section | Integral Human Development, https://migrants-refugees.va/country-profile/zambia/

Migrants/Refugees. 2022. *Migration Profile, Zimbabwe*. Vatican City: Migrants And Refugees-Integral Human Development, https://migrants-refugees.va/fr/blog/country-profile/zimbabwe/ [Accessed 14/08/2022].

Mikkelsen, B. 2005. *Methods for Development Work and Research: A New Guide for Practitioners*, SAGE.

Miller, P. 1991. *Motivation in the Workplace. Work and Organizational Psychology*. Oxford: Blackwell Publishers.

Mitchell, J. C. 1985. Towards A Situational Sociology of Wage-Labour Circulation. In Prothero, R.M. & Chapman, M. (Eds.), *Circulation in Third World Countries*. Routledge and Kegan Paul, 30–53.

Monson, T., Landau, L., Misago, J. & Polzer, T. 2010. May 2008 Violence Against Foreign Nationals in South Africa: Understanding Causes and Evaluating Responses.

Mooya, M. M. & Cloete, C. E. 2007. Informal Urban Property Markets and Poverty Alleviation: A Conceptual Framework. *Urban Studies*, 44, 147–165.

Morapedi, W. 2003. *Post Liberation Xenophobia and the African Renaissance: The Case of Undocumented Zimbabwean Immigrants into Botswana C 1995–2003*. Gaborone: History Departmental Seminar, In University of Botswana.

Moroka, T. & Tshimanga, M. 2009. Barriers to and Use of Health Care Services among Cross-Border Migrants in Botswana: Implications for Public Health. *International Journal of Migration, Health and Social Care*.

Mthembu-Salter, G., Amit, R., Gould, C. & Landau, L. B. 2014. Counting the Cost of Securitising South Africa's Immigration Regime.

Mulenga, C. L. 2003. Lusaka, Zambia. *Case Study for the United Nations Global Report on Human Settlements*.

Murray, M. J. 2011. *City of Extremes: The Spatial Politics of Johannesburg*. Duke University Press.

Mvula, L. D. 2009. Refugee Status Determination and Rights In Malawi. Refugee Studies Centre Workshop Discussion on RSD and Rights in Southern and East Africa, Uganda.

Mwakikagile, G. 2010. *Zambia: Life In An African Country*. New Africa Press.

Ndegwa, D. 2015. *Migration in Malawi: A Country Profile 2014*. International Organization for Migration.

Nejad, M. N. & Young, A. T. 2016. Want Freedom, Will Travel: Emigrant Self-Selection According to Institutional Quality. *European Journal of Political Economy*, 45, 71–84.

Neli, E., Pugliese, A. & Ray, J. 2013. *The Demographics of Global Internal Migration*. International Organization for Migration.

Nengwekhulu, R. 2008. *An Evaluation of the Nature and Role of Local Government in Post Colonial Botswana*. University of Pretoria.

Newton, K. & Van Deth, J. W. 2016. *Foundations of Comparative Politics: Democracies of the Modern World*. Cambridge University Press.

Nhema, A. G. 2002. *Democracy in Zimbabwe: From Liberation to Liberalization*. University of Zimbabwe Publications Office.

Nkhoma, B. G. 2014. Transnational Threats: The Problem of Illegal Immigration in Northern Malawi. *Southern African Peace and Security Studies*, 1, 29–43.

Norris, J. 1975. *Functions Of Ethnic Organizations*. Functions of Ethnic Organizations, 165–176.

North, D. C. 1990. *Institutions, Institutional Change and Economic Performance*. Cambridge University Press.

North, D. C. 1991. Institutions. *Journal of Economic Perspectives*, 5, 97–112.

Northcote, M. A. 2015. *Enterprising Outsiders: Livelihood Strategies of Cape Town's Forced Migrants*. University of Western Ontario.

Nshimbi, C. C. & Fioramonti, L. 2013. A Region without Borders? Policy Frameworks for Regional Labour Migration Towards South Africa. *Nshimbi, Cc & Fioramonti, L. (2013) Miworc Report*.

Nshimbi, C. C., Moyo, I. & Gumbo, T. 2018. Between Neoliberal Orthodoxy and Securitisation: Prospects and Challenges for a Borderless Southern African Community. In *Crisis, Identity and Migration in Post-Colonial Southern Africa*. Springer International Publishing AG, 167–186.

Nthanda, N. 2016. Xenophobic Attacks Erupt in Lusaka. *Iol.* https://migrants-refugees.va/country-profile/zambia/

Ntseane, D. & Mupedziswa, R. 2018. Fifty Years of Democracy: Botswana's Experience in Caring for Refugees and Displaced Persons. *International Journal of Development and Sustainability*, 7, 1408–1427.

Ntseane, D. & Solo, K. 2007. Access to Social Services for Non-Citizens and the Portability of Social Benefits within the Southern African Development Community—Botswana Country Report. Background Paper for Joint Ids/World Bank Research Project.

Ogbu, J. U. & Matute-Bianchi, M. E. 1986. Beyond Language: Social and Cultural Factors in Schooling Language Minority Students. In Perlmann, J. & Vermeulen, H. (Eds) *Bilingual Education Office.* Los Angeles, CA: Evaluation, Dissemination, and Assessment Center. Immigrants, Schooling, and Social Mobility: Does Culture Make a Difference.

Okyere, D. 2018. Economic and Social Survival Strategies of Migrants in Southern Africa: A Case Study of Ghanaian Migrants in Johannesburg, South Africa.

Oosthuizen, M. & Naidoo, P. 2004. Internal Migration to the Gauteng Province. Development Policy Research Unit. Working Paper 04/88, University of Cape Town.

Palmer, R. 1979. A History of Rhodesia. JSTOR.

Parilla, J. & Trujillo, J. 2015. South Africa's Global Gateway: Profiling the Gauteng City-Region's International Competitiveness and Connections. In *Global Cities Initiative, A Joint Project of Brookings and Jp Morgan Chase.* Global Cities Initiative.

Parsons, N. 2008. The Pipeline: Botswana's Reception of Refugees, 1956–68. *Social Dynamics*, 34, 17–32.

Pasteur, D. 1982. *The Management of Squatter Upgrading in Lusaka, Phase 2: The Transition to Maintenance and Further Development.* Development Administration Group, Institute of Local Government Studies.

Patton, M. Q. 1990. *Qualitative Evaluation and Research Methods.* Sage Publications, Inc.

Peberdy, S. 2013. Gauteng: A Province of Migrants. *Gauteng City-Region Observatory Data Brief.*

Peberdy, S. 2016. *International Migrants in Johannesburg's Informal Economy.* African Books Collective.

Polzer, T. 2010. *Migration Fact Sheet 1: Population Movements in and to South Africa.* Johannesburg: Fmsp.

Porta, J. 2012. *Just We Look for Peaceful Life: Forced Migrants in Botswana and Decision Making in the Transit Experience.* University of Toronto Scarborough.

Portes, A. & Böröcz, J. 1989. Contemporary Immigration: Theoretical Perspectives on its Determinants and Modes of Incorporation. *International Migration Review*, 23, 606–630.

Potts, D. 1986. *Urbanization in Malawi with Special Reference to the New Capital City of Lilongwe.* University of London.

Putnam, R. D. 2000. *Bowling Alone: The Collapse and Revival of American Community.* Simon and Schuster.

Qadeer, M. 1981. The Nature of Urban Land. *American Journal of Economics and Sociology*, 40, 165–182.

Republic of Zimbabwe. 2013. *Constitution of Zimbabwe Amendment (20).* Harare: https://www.constituteproject.org/constitution/zimbabwe_2013.pdf

Robertson, C. C. 1986. *Remembering Old Johannesburg*. Ad. Donker.
Ross, N. K. 2005. *Sample Design for Educational Survey Research*. International Institute for Educational Planning/Unesco, 2.
Rust, K. & Gavera, C. 2013. *Housing Finance in Africa: A Review of Some of Africa's Housing Finance Markets*. Johannesburg, SA: Center for Affordable Housing Finance in Africa.
Rutinwa, B. 2002. Asylum and Refugee Policies in Southern Africa: A Historical Perspective. In *Legal Resources Foundation, A Reference Guide to Refugee Law and Issues in Southern Africa*, 1–10.
Sachikonye, L. M. 2002. Whither Zimbabwe? Crisis & Democratisation. *Review of African Political Economy*, 29, 13–20.
SACN. 2018. *State of City Finances Report 2018*. Johannesburg: South African Cities Network (SACN).
Saunders, M., Lewis, P. & Thornhill, A. 2009. *Research Methods for Business Students*. Pearson Education.
Schäffler, A., Christopher, N., Bobbins, K., Otto, E., Nhlozi, M., De Wit, M., Van Zyl, H., Crookes, D., Gotz, G. & Trangoš, G. 2013. State of Green Infrastructure in the Gauteng City-Region. In *Gauteng City-Region Observatory (Gcro), A Partnership of the University of Johannesburg*. The University of the Witwatersrand, Johannesburg, and the Gauteng Provincial Government.
Seshamani, V. 2002. *Zambia Poverty Reduction Strategy Paper: 2002–2004*. Lusaka: Unpublished.
Seth, W. 2008. Major Urban Centres. In *Botswana and Its People*. South Africa: New Press, 44–46.
Shindondola, H. 2003. *Xenophobia in South Africa and Beyond: Some Literature for a Doctoral Research Proposal*. Johannesburg: Rand Afrikaans University.
Short, J. R. 2015. *Why Cities are a Rare Good News Story in Climate Change*. Umbc Faculty Collection.
Short, J. R. 2021. Social Inclusion in Cities. *Frontiers in Sustainable Cities*, 3, 22.
Shorten, R. 1966. *The Johannesburg Saga, John R*. Johannesburg: Shorten Propriety Limited.
Skeldon, R. 2013. *Global Migration: Demographic Aspects and Its Relevance for Development*. New York: United Nations.
Smit, R. & Rugunanan, P. 2014. From Precarious Lives to Precarious Work: The Dilemma Facing Refugees in Gauteng, South Africa. *South African Review of Sociology*, 45, 4–26.
Soja, E. 2010. *Globalization and Community: Seeking Spatial Justice*. Minneapolis, US: University of Minnesota Press. Retrieved From http://www.ebrary.com
Southall, R. 1984. Botswana as a Host Country for Refugees. *Journal of Commonwealth & Comparative Politics*, 22, 151–179.
Speziale, H. S., Streubert, H. J. & Carpenter, D. R. 2011. *Qualitative Research in Nursing: Advancing the Humanistic Imperative*. Lippincott Williams & Wilkins.
Stats SA 2015. Census 2011: Migration Dynamics in South Africa. In *Statistics South Africa*. Pretoria: Statistics South Africa.
Storie, M. 2014. Changes in the Natural Landscape. In *Changing Space Changing City: Johannesburg after Apartheid*. Johannesburg: Wits University Press, 137–153.
Suphanchaimat, R., Kantamaturapoj, K., Putthasri, W. & Prakongsai, P. 2015. Challenges in the Provision of Healthcare Services for Migrants: A Systematic Review through Providers' Lens. *BMC Health Services Research*, 15, 1–14.

Sustainability for All. 2019. https://www.activesustainability.com/Constructionand-Urban-Development/Cities-Communities-Sustainable/?_Adin=02021864894

Symonds, F. A. 1953. *The Johannesburg Story*, F. Muller.

Taruvinga, R., Hölscher, D. & Lombard, A. 2021. A Critical Ethics of Care Perspective on Refugee Income Generation: Towards Sustainable Policy and Practice in Zimbabwe's Tongogara Camp. *Ethics and Social Welfare*, 15, 36–51.

Teixeira, C. & Halliday, B. 2010. Introduction: Immigration, Housing and Homelessness. *Canadian Issues*, 3.

Tevera, D. & Zinyama, L. 2002. *Zimbabweans Who Move: Perspectives on International Migration in Zimbabwe*. Idasa and Queens University.

Timngum, D. A. 2001. *Socio-Economic Experiences of Francophone and Anglophone Refugees in South Africa: A Case Study of Cameroonian Urban Refugees in Johannesburg*. University of the Witwatersrand.

Tomlinson, R., Beauregard, R., Bremmer, L. & Mangcu, X. 2003. *Emerging Johannesburg*. New York and London: Routledge.

Turok, I. 2012. *Urbanisation and Development in South Africa: Economic Imperatives, Spatial Distortions and Strategic Responses*. Human Settlements Group, International Institute For Environment And ....

UN DESA. 2004. *Proceedings of the Third Coordination Meeting on International Migration*. New York. www.un.org/en/development/desa/population/migration/events/coordination/4/docs/Report_Third_Coordinationmeeting.pdf

UN DESA. 2013. International Migration 2013 Wallchart. www.un.org/en/development/desa/population/migration/publications/wallchart/docs/wallchart2013.pdf

UN-Habitat. 2010. Malawi Urban Housing Sector Profile. 131 P. https://issuu.com/unhabitat/docs/malawi_urban_housing_sector_profile

UN-Habitat. 2018. *Tracking Progress Towards Inclusive*. Safe, Resilient and Sustainable Cities and Human Settlements.

UN-Habitat. 2018a. *Migration And Inclusive Cities. A Guide For Arab City Leaders*. United Nations Human Settlements Programme.

UN-Habitat. 2018b. *Tracking Progress Towards Inclusive*. Safe, Resilient and Sustainable Cities and Human Settlements.

UNHCR. 2000. *UNHCR Global Report 1999*. Geneva: United Nations High Commission for Refugees (UNHCR)

UNHCR. 2006. *Country Operations Plan (Botswana)*. Gaborone.

UNCHR. 2012. *UNHCR Global Report 2011*. Geneva: United Nations High Commission for Refugees (UNHCR).

UNHCR. 2014. *UNHCR Global Report- Southern Africa Sub Regional Overview*. Geneva.

UNHCR. 2016. UNHCR Operation in Zimbabwe Fact Sheet. www.unhcr.org

UNHCR. 2018. *Zimbabwe* [Online]. http://reporting.unhcr.org/Node/10232 [Accessed 2018].

UNHCR. 2020. Zimbabwe. *Fact Sheet*. https://Reporting.Unhcr.org/Sites/Default/Files/Unhcr%20mozambique%20fact%20sheet%20december%202020.Pdf

UNHCR. 2021. *Zimbabwe: Fact Sheet* [Online]. http://reporting.unhcr.org/Zimbabwe [Accessed 21/02/2022].

UNHCR. 2022. *UNHCR's Protection Chief Concludes Three-Day Visit to Zimbabwe* [Online]. https://www.unhcr.org/afr/news/press/2022/1/61ee72634/unhcrs-protection-chief-concludes-three-day-visit-to-zimbabwe.html [Accessed 07/07/2022].

UNHCR & WFP. 2014. *Zimbabwe - UNHCR/WFP Joint Assessment Mission Report: Tongogara Refugee Camp, September 2014* [Online]. https://www.wfp.org/Publications/Zimbabwe-Unhcr-Wfp-Joint-Assessment-Mission-Tongogara-Refugee-Camp-September-2014 [Accessed 14/08/2022].

US Department of State. 2013. *Human Rights Report, Botswana*. Bureau of Democracy, Human Rights and Labour.

US Department of State. 2020. *2020 Trafficking in Persons Report: Zimbabwe* [Online]. https://www.state.gov/reports/2020-trafficking-in-persons-report/zimbabwe/ [Accessed 14/08/2022].

Vandeyar, S. 2010. Educational and Socio-Cultural Experiences of Immigrant Students in South African Schools. *Education Inquiry*, 1, 347–365.

Vearey, J. & Nunez, L. 2010. Migration and Health in South Africa. *Background Paper and Report on the National Consultation on Migration and Health in South Africa, Midrand, 22nd–23rd April*.

Walls, H. L., Vearey, J., Smith, R. D., Hanefeld, J., Modisenyane, M., Chetty-Makkan, C. M. & Charalambous, S. 2016. Understanding Healthcare and Population Mobility in Southern Africa: The Case of South Africa. *South African Medical Journal*, 106, 14–15.

WEF 2017. Migration and Its Impact on Cities. *World Economic Forum in Collaboration with PWC*.

West, G. 2018. *Scale: The Universal Laws of Life, Growth, and Death in Organisms, Cities, and Companies*. Penguin.

Whitehouse, B. 2012. *Migrants And Strangers In An African City: Exile, Dignity, Belonging*. Indiana University Press.

WHO 2010. Health of Migrants: The Way Forward: Report of a Global Consultation, Madrid, Spain, 3–5 March 2010.

Willet, S. M., Schmid, J. & Hudson, D. J. 2015. *Voices of Kagisong: History of a Refugee Programme in Botswana*. Bay Publishing.

Williams, G. J. 1984. *The Peugeot Guide to Lusaka*. Zambia Geographical Association.

Williamson, O. E. 2000. The New Institutional Economics: Taking Stock, Looking Ahead. *Journal of Economic Literature*, 38, 595–613.

World Bank. 2019. *Structural Transformation Can Turn Cities into Engines of Prosperity*, April 17. Available At: https://www.worldbank.org/en/news/feature/2019/04/17/structural-transformation-can-turn-cities-into-engines-of-prosperity

Xie, X. 2018. Main Points of the Theory of Intersubjectivity. 4th International Conference on Humanities and Social Science Research (Ichssr 2018). Atlantis Press, 149–153.

Zhou, M. 1997. Segmented Assimilation: Issues, Controversies, and Recent Research on the New Second Generation. *International Migration Review*, 31, 975–1008.

Zulu, E. 2005. Interpreting Exodus from the Perspective of Ngoni Narratives Concerning Origins: Appropiating Exodus in Africa. *Scriptura: Journal for Contextual Hermeneutics in Southern Africa*, 90, 892–898.

3 Immigrants and the City
A Conceptual Framework

3.1 Introduction

The modern world is at war in a modest way, largely caused by the increase in migration. In reality, it is generally accepted that migration is a universal phenomenon and it is here to stay. Governments around the world respond to the influx of migrants in various ways, which in most cases involves trying to safeguard the interests of their countries yet being accommodative of immigrants. The process of balancing the two has always been a source of attrition between immigrants and indigenous people, but more so, a grey area when it comes to government policies.

This chapter revisits this tenacious issue and conceptualises it from three perspectives which provide a holistic platform with the intention of understanding inclusivity in cities:

- Migration and the international economic system;
- The institutional perspective of migration; and
- Acculturation – a four-model approach.

This diverse approach is appropriate in the context of this book because it provides a firm foundation to analyse the city from a holistic approach where the interplay of economics and migration laws are expressed spatially at the city level. The city, in the essence of this book, emerges as the spatial point of analysis since the built form cannot be separated from the movement of people and livelihoods irrespective of the push factors of migrants. Hence, most urban economic hubs are major recipients of such migrants. However, in the context of developing countries, there are conceptual challenges to analysing and understanding immigrants because of the excess numbers of undocumented immigrants who operate in a purely informal environment.

3.2 Migration and the International Economic System

The world economic system is not a surprise entry into the discourse of migration and the city. This should be understood from a common perspective where human resources and the physical concentration of capital tend to

DOI: 10.4324/9781003184508-3

form the basis upon which livelihoods thrive. This is a purely Marxist perspective of understanding migration because it fits immigrants in an economic framework where their survival is dictated by the host economic system. Amankwaa (1995), in the analysis of international migration in less developed countries, observed that the context, size, composition and predisposition of people to move to another country is largely driven by the international economic system. This is understood from the fact that contemporary richer nations have at some time in the development process exploited poorer nations of the world – a phenomenon which led to underdevelopment.

Haralambos and Holborn (1995) take a historical path on this issue by analysing how European international capital was used to exploit Third World resources (such as labour and raw materials). This was a full-fledged stage of exploitation where colonies were not given the opportunity to develop, let alone develop industries to sustain themselves. This uneven development left most Third World countries impoverished. They were simply used as a reserve army of labour whose benefits were either channelled to developed countries or, alternatively, used to build prosperous colonial cities which were a reserve of the few settlers. The creation of such dual economies, be they global, regional or within countries, has given rise to endless migration as people seek opportunities where there is economic prosperity.

Harvey (2020), writing in the context of imperial capitalism, provides further insights into this perspective by analysing how capital was used to transform colonial environments (in the form of built environments such as urban centres) and concentrated social power in such spaces. The ability to establish private property, transport and communication systems (among others) through repression, law and education shows the power of capitalist imperialism that was used to run speculative investments. In the process, immigrants were also shipped in to augment local labour reserves. Ironically, the trend is still persisting, though voluntarily, as people continue to move globally in search of opportunities, safety and peace.

What is interesting about this approach to immigrants is its spatial impact on the city. Kosack (in Haralambos and Holborn (1995:672)), writing in the context of developed countries (i.e. France, Germany, Britain and Switzerland), noted that immigrants normally take 'a subordinate position on the labour market', which in turn contributes to spatial exclusion. As summarised in Box 3.1, most immigrants are employed as manual labourers whose wages are low. This impacts negatively on their survival in the sense that they cannot afford decent services such as housing.

It seems that the process of enriching developed countries continues unabated as immigrants are subjected to prejudice in receiving countries. The main beneficiaries are the ruling class, the rich and the powerful who continue to benefit from cheap labour and continue the discrimination of immigrants. However, even in the context of developing countries, the exploitation of immigrants for labour is a well-documented discourse. Chipungu and Magidimisha (2020) and Magidimisha-Chipungu and Chipungu (2021) capture this narrative vividly in the context of Southern Africa (specifically in

> **Box 3.1 The Spatial Impact of Migration and International Economic System on the Inclusivity of Cities**
>
> General observations show that immigrants at the city level are regarded as second-class citizens. They:
>
> - Are commonly found in run-down areas of the city;
> - Occupy dilapidated housing in poor areas of the city;
> - Lack educational opportunities;
> - Suffer widespread prejudice and discrimination from local authorities;
> - Suffer widespread discrimination from indigenous people;
> - Occupy subordinate position on the labour market; and
> - Are used as scapegoats for crime.
>
> Source: Author from various sources (2022).

Zimbabwe and South Africa) from both a colonial and a post-colonial perspective. Hence its perseverance and persistence in post-colonial countries is indeed a cause for concern despite the fact that there are other forces at work.

3.3 The New Institutional Approach

One approach which can be used to analyse inclusivity among immigrants in cities is through the new institutional approach. An institutional approach to understanding immigrants and the issue of inclusivity within cities builds upon and extends the existing neoclassic theories to the explanation of social problems. It draws its major strength from the premise of including institutions in the inquiry of social problems. Qadeer (1981) argues that the institutional approach perceives developmental outcomes as products of a series of public and private decisions embedded in the social and political institutions from which they cannot be separated conceptually. It should be remembered that neo-classic theories envisaged institutions as unnecessary and believed that development could be explained within the framework of instrumental rationality (Mooya and Cloete, 2007; North, 1991). However, it is a reality that cannot be denied that development together with its associated consequences operates in an institutional environment.

North (1991) defines institutions as humanely constructed constraints that structure and define the interaction of human beings. He argues that the human environment is made up of complexities of problems that at times can defy economic rationality. Hence rules and formal hierarchies are put in place to shape actions and expectations (North, 1990). He further asserts that institutions come in the form of formal rules, informal constraints and enforcement tools. These institutions operate through an array of socio-economic and political organisations. Organisations are therefore groups of individuals

Table 3.1 Levels of Social Analysis

Level	Purpose
1	Embeddedness
	Informal institutions, e.g. culture, tradition, norms
2	Institutional environment
	Formal rules of the game, e.g. bureaucracy, polity
3	Governance: *Play of the game*
4	Resource allocation

Source: Williamson, 2000:595–613.

bound by a common purpose with the aim of achieving certain objectives and they include political, economic and social bodies (North, 1991).

These institutions do not operate in their individual capacity, but they are interdependent. It is this element of interconnectedness that eventually results in the formation of organisational structures through which transactions are effected (North, 1991). Williamson (2000) clearly portrays this element of interdependence diagrammatically in his schematic four Levels of Social Analysis, as shown in Table 3.1. He noted that although there is interaction between the various levels of social analysis, the higher levels impose constraints on the lower levels immediately below them.

Institutions as determinants of migration provide a complex point of analysis. They are, on the other hand, an important point of entry into this discourse where the immigrant–host relationship is at the centre of discussion. Migration, as a process, is not simply a movement of people but a complex phenomenon governed by international, regional and national institutions. These come in different forms such as international treaties, protocols, legislations policies and organisations (both at the international and the national levels) whose mandate is to facilitate the movement of people within and across national borders. For this reason, institutions cut across all sectors of migration and the relative position of immigrants under the spotlight in this book.

As already observed under the international economic systems earlier, migrants are attracted by economic freedom. As Nejad and Young (2016) postulate, migrants are more likely to migrate to countries with sound currencies, less burdensome regulations, and stronger property rights and legal systems. This, in essence, speaks to strong labour market institutions which provide immigrants with freedom of entry and a platform to bargain for their continual survival. Accommodative labour economic institutions have a positive impact on migrants irrespective of their status, i.e. be they legal or illegal migrants (Geis et al., 2013). However, liberal labour institutions, on the other hand, have come under the spotlight in countries where unemployment is high and competition for opportunities is stiff between locals and immigrants. Hence, the city's level of inclusivity can be measured by analysing labour economic institutions in terms of how they are accommodative of immigrants.

In the same line of resonance, political institutions are also major determinants when it comes to attracting and stabilising migrants within their own boundaries. Bertocchi and Strozzi (2008), for instance, argue that the level of democracy in a given country is a strong predictor when it comes to attracting migrants. They further noted that what attracts migrants are levels of democracy (such as citizenship laws), immigration policies, policing mechanism as well as access to social and public facilities. Hence political institutions are key determinants of political freedom, which in turn impacts on the freedom of migrants (Ashby, 2010).

In summary, it can be argued that the interplay of various institutional mechanisms does not only determine the inflow of migrants into a particular country but also provides insights into the level of acceptability of immigrants in society, thereby determining their willingness to stay. Hence, the level of inclusivity in any city is also largely a function of various institutions in terms of their response to immigrant issues.

3.4 Acculturation – The Four Models Approach

At the centre of migration are dynamics that involve immigrants and host countries. The immigrant–host relationship is a complex union which is fluid, undefined and at times institutionalised (Haralambos and Holborn, 1995). The complexity of the situation emanates from the socio-economic and cultural dimensions of immigrants, which in most cases is at variant with the host society. This is further aggravated by divergent intentions which both groups have in terms of their expectations and intentions in life. This, in essence, is a 'forced-marriage' type of existence where social relations, irrespective of backgrounds, are supposed to work harmoniously through voluntary means. Hence, what transpires in the process can lead to acceptance or rejection of one's culture depending on the immigrant–host relationship. It is from such a background that an in-depth understanding of the immigrant–host relationship is provided by the psychological theory of acculturation, which, in principle, emanates from the need to balance two cultures while adapting to the prevailing one. Psychologically, acculturation as a process is simply associated with individuals who try to adapt, acquire and adjust to new cultural environments.

This, in essence, is a gradual process which involves individuals trying to fit into the new environment while still holding on to their original cultural values and norms (Berry, 1997). The process of acculturation is subtle since it operates at different levels in society and responds to various factors (such as cultural, institutional and environmental – among others), which can impend or smoothly facilitate expected changes. These significant changes which individuals try to adapt to can be comprehensive – a situation which has prompted psychologists to label acculturation as the second-culture learning.

Various authors (Berry, 1997; Zhou, 1997; Berry et al., 2006) provide further insights into the immigrant–host relationship by arguing that acculturation is not a smooth process, which can lead to minorities (being immigrants) adopting or rejecting native cultures. Using the fourfold bilinear model, Zhou (1997), for instance, provides a framework for acculturation that revolves around four strategies. These strategies stem from two critical questions where immigrants have to weigh the value of maintaining their own identity or adopting those of the host society. On the basis of the response, immigrants can be assimilated, separated, integrated or marginalised in the host country.

- *Assimilation:* This occurs when individuals respond positively to the cultural norms of the host over their original culture. The whole relationship hinges on adaptation where the 'the hosts and immigrants adapt to living together' (Haralambos and Holborn, 1995:669). This is illustrated by Haralambos and Holborn (1995) in the case of London, where the immigrant–host relations were more harmonious and therefore provided a platform upon which both parties were more accommodative of each other to such an extent that elements of integration and assimilation emerged. However, this is not a smooth process since, at times, it requires government intervention to make the relationship work.
- *Separation:* When immigrants decide to reject the dominant culture of the host, separation is the end result. Humans, by nature, are selfish – in this instance, the need to preserve their own culture will be the dominant factor. This is evidenced spatially where ethnic enclaves emerge or are created depending on societal dynamics such as resistance or discrimination by the host society.
- *Integration:* This scenario emerges when immigrants take the middle of the road position, i.e. on the one hand, they opt to adopt the cultural norms of the host society, while on the other hand, they continue maintaining cultural values of their origin. It is this strategy which to a large extent is associated with bi-culturalism, which, in essence, manifests in the form of a dual identity (i.e. both of immigrant's origin and those of the host society).
- *Marginalisation:* This is an extreme position where immigrants reject both their cultural values and those of the host society. In essence, marginalisation leans more towards social exclusion since the marginalised are perceived as a sub-group without identity to either the immigrants or the hosts. As a process, marginalisation comes with its own package of exclusion among which is lack of power, resources, status and recognition (among others).

However, it must be noted that in as much as these strategies are very clear, their application in reality is driven by many factors. There are both personal (or private) and societal (public) factors that influence the whole process of acculturation. On the other hand, the reality on the ground is that immigrants, who in other words 'intrude' in a new environment, are expected to do more in the adaption process than the host community. This stems out of the fact that they have 'to undergo changes' in their lifestyles while the host take a 'passive' and unconscious position in the adaptation process. In any case, hosts do not have anything to lose – hence the whole process of adaptation is a one-sided process where immigrants have to re-socialise in order to be acceptable in this new society.

3.5 Inclusivity and Migrants – A Conceptual Summary

Cities emerge as melting pots of society where multi-culturalism is supposed to be thriving, benchmarked by healthy, harmonious relationships achieved through assimilation. The three conceptual anchors discussed in this chapter attempted to drive the discourse of immigrant–host relationships towards a more reasonable platform upon which inclusivity in modern-day cities can be built. They provided an economic-institutional-based discourse which goes beyond the perception of the city in its physical perspective. This is an integral turning point in the narrative of cities because the physical form is moulded by human intervention. It is therefore this human dimension which should be understood and therefore forms the basis upon which one can interrogate inclusivity in urban space. Hence, the whole chapter can be summarised in three critical perspectives emanating from the theoretical underpinnings provided – viz. cities as economic organisms; cities as regulated platforms; and the human dimension of cities. The success or failure of inclusivity, to a large extend, is a result of the interplay of these perspectives.

Cities as economic organisms: The perception of cities as centres of production is a universal viewpoint which elevates cities to be economic centres. Their growth is largely driven by resource endowment coupled with international capital–created magnetic forces that attract more capital, which is, in turn, invested in various physical dimensions such as infrastructure, commercial, industrial and social developments. It is for this reason that cities have always remained centres of attraction as people from diverse backgrounds and nationalities descended upon them in search of economic opportunities. The concentration of people from diverse backgrounds has always been a source of attrition and this is evidenced by the physical manifestation of the city. It is in such a situation that immigrants (especially those who are not documented) are relegated to the position of underdogs. This is not by accident, especially in an environment where capital accumulation by those who own the means of production largely depends on cheap labour.

Cities as regulated platforms: The spatial organisation of the city is not by chance but is a product of rules and regulations that dictate the use of space and also the behaviour of the users of such spaces. Hence economic production alluded to in the above paragraph operates in such a regulated environment. It is on this platform that policies and legislations pertaining to immigrants are provided in a bid to achieve inclusivity. Yet in the same line of resonance, it is not by chance that extra-legal developments emerge – in essence, they are products of regulated environments where partakers try to survive outside such a system. While the regulated environment provides a platform for smooth functionality, the illegal environment provides platforms for survival. Hence, most immigrants (especially in developing countries, which in most cases are home to these illegal environments) have learnt to exist outside the regulated environments in their search for opportunities.

The human dimension of cities: The human dimension of cities is largely driven by the economic status coupled with the institutional framework.

Access to economic opportunities by immigrants is not only a function of their availability but also depends on the accommodative nature of both the working and the living domains of the host environments. These emanate from the prevailing institutional frameworks of the host society, which may or may not allow accommodation of immigrants. This, in turn, has implications for the nature of relationships between the hosts and the immigrants (leading to assimilation and integration). Therefore, the acceptance of immigrants in any host society is a complex process which has a direct impact on the inclusivity of the city, both demographically and spatially. In essence, the immigrant–host relationship is a transformative process in its own right which impacts on both the hosts and the immigrants.

In conclusion, the conceptual perspectives discussed in this chapter provide the basis upon which the inclusivity of cities of Southern Africa will be analysed in the succeeding chapters. It should be reiterated emphatically here that this book does not attempt to re-invent a new wheel – but rather, to contribute towards the narrative on inclusivity and immigrants in the cities of Southern Africa based on existing institutional frameworks prevailing in these countries. However, foresight should not be lost of the holistic dimension of immigrant inclusivity laid down in Chapter 1 which actually sets out the agenda for the whole book.

Bibliography

Amankwaa, A. A. 1995. The World Economic System and International Migration in Less Developed Countries: An Ecological Approach. *International Migration (Geneva, Switzerland)*, 33, 93–114.

Ashby, N. J. 2010. Freedom and International Migration. *Southern Economic Journal*, 77, 49–62.

Berry, J. W. 1997. Immigration, Acculturation, and Adaptation. *Applied Psychology*, 46, 5–34.

Berry, J. W., Phinney, J. S., Sam, D. L. & Vedder, P. 2006. Immigrant Youth: Acculturation, Identity, and Adaptation. *Applied Psychology*, 55–303–332.

Bertocchi, G. & Strozzi, C. 2008. International Migration and the Role of Institutions. *Public Choice*, 137, 81–102.

Chipungu, L. & Magidimisha, H. H. 2020. *Housing in the Aftermath of the Fast Track Land Reform Programme in Zimbabwe*. Routledge.

Geis, W., Uebelmesser, S. & Werding, M. 2013. How Do Migrants Choose Their Destination Country? An Analysis of Institutional Determinants. *Review of International Economics*, 21, 825–840.

Haralambos, M. & Holborn, M. 1995. *Sociology. Themes and Perspective* 4th Edition. London: Collins Educational.

Harvey, D. 2020. *The Condition of Postmodernity. The New Social Theory Reader*. Routledge.

Magidimisha-Chipungu, H. H. & Chipungu, L. 2021. *Urban Inclusivity in Southern Africa*, Springer.

Mooya, M. M. & Cloete, C. E. 2007. Informal Urban Property Markets and Poverty Alleviation: A Conceptual Framework. *Urban Studies*, 44, 147–165.

Nejad, M. N. & Young, A. T. 2016. Want Freedom, Will Travel: Emigrant Self-Selection According to Institutional Quality. *European Journal of Political Economy*, 45, 71–84.

North, D. C. 1990. *Institutions, Institutional Change and Economic Performance*, Cambridge University Press.

North, D. C. 1991. Institutions. *Journal of Economic Perspectives*, 5, 97–112.

Qadeer, M. 1981. The Nature of Urban Land. *American Journal of Economics and Sociology*, 40, 165–182.

Williamson, O. E. 2000. The New Institutional Economics: Taking Stock, Looking Ahead. *Journal of Economic Literature*, 38, 595–613.

Zhou, M. 1997. Segmented Assimilation: Issues, Controversies, and Recent Research on the New Second Generation. *International Migration Review*, 31, 975–1008.

4 The Southern African Region in a Historical Perspective

4.1 Introduction

Writing a book straddling over six countries is not an easy feat. However, the search for critical insight into contemporary societal dynamics with crosscutting factors that impact across borders of sovereign countries fuelled the need to explore their depth. The issue of migration within the Southern African Development Community (SADC) Region is not a new phenomenon. However, its perpetuation over time calls for the need to continue monitoring its dynamics with the intention of establishing constructive ameliorative platforms that can contribute to shared solutions. The intensity of migration over the years has soared to levels which some sovereign countries are battling to contend with – yet solutions seem to be floating away beyond any thinkable horizons.

A lot has been written about migration in Southern Africa, yet there seems to be no solution to its dynamics. In our previous book on migration, *Crisis, Identity and Migration in Post-Colonial Southern Africa* in 2018, we interrogated various thematic issues that revolved around *migration patterns*, *post-colonial governance* and *citizenship*. This book takes the narrative further by exploring migration from the position of immigrants. It stems out of the premises that immigrants are a vulnerable group of people who are marginalised in host countries. Of late, the search for economic freedom, especially in the context of the SADC Region, is complicating the equation. The composition of migrants is constantly changing as other vulnerable members of society (such as women, children and the elderly) are becoming part of active migrants. It is from this perspective that this book interrogates the extent to which immigrants are assimilated into the host country through various inclusionary policies.

4.2 The Preamble to Migration Dynamics in Southern Africa – The Pre-colonial Stage

The history of Southern Africa is one of displacements and this is depicted throughout the pre-colonial period, the colonial period and the post-colonial period. These movements in 'migratory societies', as Lesthaeghe (1989) refers

DOI: 10.4324/9781003184508-4

to them, have pervasive social, political and economic impacts on contemporary states. It is essential to reflect on the pre-colonial stage of Southern African countries since some of the dynamics of the traditional societies (African kingdoms) have had an overwhelming impact on contemporary migration trends. Two major pre-colonial historical developments – the *Mfecane* (forced migration) and the Great Trek – need a brief overview in this context.

The *Mfecane*, which occurred between 1750 and 1835, originated in South Africa's KwaZulu-Natal Region and its impact spanned across most countries in Southern Africa (mainly, modern-day Mozambique, Malawi, Zimbabwe, Zambia, Botswana, Lesotho, Swaziland and Tanzania). Documented evidence from many historians such as Kowet (1978), Magubane (1975) and Mitchell (1985), among others, concur that the forced migrations which eclipsed most of Southern Africa were mainly generated by drought, conflict over land, grain, cattle, water and trade. At the centre of these dynamics were the political dynamics of state building by various kingdoms, mainly Shaka, the Zulu King. Driven by all these forces, Shaka's most trusted lieutenants and the members of his inner circle of the Nguni tribal groupings revolted and broke away from the Zulu Kingdom (Zulu, 2005). These triggered years of forced migration as various tribal groups fled from Shaka – a process which led to the formation of new Kingdoms and current-day states, as shown in Table 4.1.

Towards the end of the *Mfecane* (around 1835), another wave of movement, triggered by the encroachment of British rule around the Cape, came in the form of the Great Trek. The Boers, who were mainly Dutch-speaking colonists, were mainly farmers who loathed falling under the British rule. Driven by the 'empty land myth' to feed their farming ego, they migrated north, where they came into conflict with various African Kingdoms some of

Table 4.1 Outcome of the *Mfecane*

The Ndebele People
The Ndebele under the leadership of Mzilikazi fled across the Limpopo River and settled in current-day southern Zimbabwe in Bulawayo after crushing the Mwene-Mutapa Kingdom.
The Sotho People
Led by Moshoeshoe, the ba-Sotho people fled and established the current-day Kingdom of Lesotho, where their kingdom eventually became a British Protectorate during the colonial period.
The Ngwane People
Under the leadership of Sobhuza, the Ngwane people moved to current-day Swaziland, where they eventually sort British protection.
The Ngoni People
Zwangendaba led his people through modern-day Mozambique, Malawi, Northern Zambia and Southern Tanzania, where his people eventually established strongholds, though scattered throughout these countries.

Source: Author from various sources (2022).

which were fleeing from Shaka, the Zulu King (such as the Ndebele under Mzilikazi) (South African History online, 2019).[1] Using their military prowess, land and labour became their loot as they fought fearless battles with African Kingdoms. They eventually established their strongholds around northern and north-western South Africa. The search for political and economic autonomy led to the founding of the Orange Free State and Transvaal, which were internationally recognised self-governing republics until they were annexed by the British in 1910.

It should be noted that the brief outline of historical pre-colonial migrations sketched in the preceding section simply gives an overview of the forced migrations which contributed to shaping contemporary settlements. In the same context, it should be borne in mind the occurrings painted here were more complex and pervasive and their effect were long-lasting in the periods in question. Hence, the formation of new states (of Lesotho and Swaziland) and the creation of farms (through the Great Trek) are among the key pointers of the political, economic and social impacts of these pre-colonial migrations. It is therefore not surprising that most people in the region still look back to the social ties of past years as they search for livelihoods and political asylum. The people of Botswana, Swaziland and Lesotho, as well as the Ndebele of southern Zimbabwe, still seek their social ties in South Africa – a factor that explains their large presence in the economic fields (such as mines, farms and industries) of South Africa. On the other hand, the formation of the Federation (of Rhodesia and Nyasaland) and the popular *Wenla* (colonial labour contracts in South Africa) were a further consolidation of these political, social and economy dynamics whose impact is still resurfacing today through similar driving forces (as seen through Mozambique's civil war and Zimbabwe's land reform programme). Hence, the emerging dimensions, patterns and processes of migration in Southern Africa must be analysed with these anecdotes in mind, as summarised in Box 4.1.

Box 4.1 Impact of the Pre-colonial Migrations

- ***State Formation:*** The birth of modern Lesotho and Eswatini which are sovereign states of Southern Africa and indispensable cogs in the discourse of immigration, as discussed in the following sections. Similarly, the Boer states of Transvaal and Orange Free States have equally social and economic impact on modern's cities' immigration discourse.
- ***Labour Immigrants:*** The period saw the emergence of labour migrants due to the onset of the modern economy. The use of African blacks by Boers and the importation of indentured Indians during the same period is not a coincidence but an elaborate beginning of labour migrants on farms and later on mines.

- **The Class System:** The emergence of centres of production (mines, towns and farms) also saw the beginning of social stratification and discrimination driven by economic, political and social forces. The Ndebele Kingdom, despite having a proportion of outcasts in the form of *ama-Hole* (slaves, below the *Enhla and Zansi* – which were superior tribal groupings in that order), was the most successful kingdom with a fusion and assimilation of different tribal groupings (such as the Venda, Shona, Nguni, Tswana and Sotho). Similarly, this became more elaborate at centres of production, where African blacks occupied the lowest level, followed by Indians, with whites being at the helm.

Sources: Author from different sources (2022).

Therefore, it will be an oversight to ignore these historical developments which had a major impact on contemporary cities in Southern African. For instance, how do we explain the high levels of assimilation and integration of Indians, Sotho and Tswana immigrants among South African societies in cities and the continual utter rejection of Zimbabwean, Mozambicans and Nigerians?

4.3 Labour Migrants in Southern Africa

Similarly, immigration within the Southern African region during the historical colonial state-building era can equally be understood from the dynamics of labour migration. At the centre of state building was access to land, minerals and labour control, which, in essence, formed the basis for capital accumulation through hegemonic powers. Writing in the context of Lesotho and Eswatini, Chipungu (2018) observes that migration dynamics stemmed out of the need to provide labour requirements for the Europeans plantations that were springing up. For indigenous people, working on these European plantations was not out of choice, but a forced decision which was a response to racial administrative and control mechanisms coupled with intrusive taxation (such as hut tax, tribal levies and fines) imposed on them by the British colonial settlers (Lesthaeghe, 1989; Kowet, 1978). However, these plantations were not enough to absorb 'pools of free indigenous free labour' which were looking for survival in the new modern economy. The situation was aggravated by the fact that the few mineral deposits discovered on 'European land' were inadequate to absorb extra labour. However, the solution to this problem was partially resolved in 1886 when gold and diamond deposits were discovered in Witwatersrand – South Africa (Crush, 2008). This signalled the beginning of migrant labourers as migrants from the region and abroad travelled to South Africa for work on mines, farms and into the emerging modern industrial economy.

South Africa, with its vast gold and diamond reserves, coupled with farms and emerging industries, became the major hub for migrants in the region. Observations by Crush (2008) shows that labour migrants came in two forms. On the one hand were mine workers whose movement was formalised and regulated by the South African Chamber of Mines and whose operations went beyond the South African borders (i.e. Lesotho, Swaziland, Mozambique and Botswana) in search of labourers (Trimikliniotis et al., 2008). Through the Rand Native Labour Association (RNLA), Witwatersrand Native Labour Association (WNLA corrupted to '*Wenela*' by indigenous migrants) workers were supported by their home governments through signing of bilateral agreements which entrenched the contract labour system established between 1890 and 1920. On the other hand, there were different kinds of informal movements of undocumented immigrants. Most of these clandestine movements ended up on commercial farms where the labourers were absorbed as farm workers (Crush, 2008). It is important to note that the contract labour system was highly institutionalised and its influence extended into countries such as Malawi, Zambia and Zimbabwe (discussed in the next section under the Federation). An extract of statistics by Lesthaeghe (1989 quoting Kowet, 1978) shows that there has been a gradual increase in migrant labourers from 1911 to 1975. This argument is supported by Crush (2008) who noted that this increase continued up to the post-democracy period as witnessed in mines where non-South African workers increased from 46% (in 1990) to 60% (in 1999) (Table 4.2).

Therefore, in as much as there was migration in the region, the movement was mainly one-sided into South Africa. The manufacturing and mining industrial activities of South Africa were built on the exploitation of cheap labour. The adoption of the Apartheid system based on the policy of separate development worsened the status of labour immigrants. In essence, Africans were not considered as immigrants – rather, they were contract workers whose residence was supposed to be temporary as per the bilateral agreements between the apartheid regime and the neighbouring countries of Mozambique, Lesotho and Swaziland (Crush, 2008; Lesthaeghe, 1989).

Table 4.2 Migrant Labour from Botswana, Lesotho and Swaziland to South Africa

Labour Migrants in (000s)			
Date	Botswana	Lesotho	Swaziland
1911	2.6	25.0	8.5
1921	3.3	47.0	6.0
1936	10.4	101.0	9.6
1946	NA	127.0	8.1
1956	NA	155.0	11.7
1966	45.0	117.0	19.2
1975	60.0	200.0	18.0

Source: Kowet (1978).

For this reason, labour migrants were mostly adults without their family members. Officially, the term 'immigrant' only applied to those people who could be assimilated into the white population – these being white migrants mainly from abroad. Those who could not fit into this system, especially in urban areas, were considered illegal migrants and were either excluded by deportation or forcefully transferred to commercial farms. It is therefore not surprising that South African cities were exclusive environments meant for the white supremacists. Even workers in the service and manufacturing sectors within urban areas maintained a 'two- legged existence' since they were supposed to leave urban areas when they retired from employment (Chipungu and Magidimisha, 2020). This was further emphasised in the level of services provided, such as in housing where hostels which were meant for male workers only became the main form of housing. Where family accommodation was provided, it came in the form of tied-housing in what were termed as 'black-spots', i.e. secluded housing environments for blacks, such as townships (Magidimisha and Chipungu, 2019).

Beyond labour migrants, the South African Apartheid Regime, due to its exclusionary policies, was not a favourite destination for refugees fleeing civil conflicts in the region. Most refugees sort asylum in Botswana, Mozambique and Malawi. Ironically, South Africa itself generated refugees who fled into neighbouring countries. However, in the late 1980s, the influx of refugees from Mozambique was limited to the border areas of the country where they were eventually integrated into the South African community – a situation which began to brew attrition between the host and the immigrant population in the post-apartheid period.

4.4 The Federation of Rhodesia and Nyasaland (1953–1963)

The formation of the Federation of Rhodesia and Nyasaland (i.e. Zambia, Zimbabwe and Malawi) from 1953 to 1963 was formulated on the need to build a strong economic union. There are two major drivers that led to the formation of this federation. Economically, it was built on the belief that creating a strong economic hub based on three countries under British colonial power would be beneficial as opposed to individual growth. This was emphasised on the backdrop of the vast copper mines of Zambia coupled with the gold mines and industrial hubs of Zimbabwe, despite the relative lack of mineral deposits in Malawi (Palmer, 1979). Politically, it was built on the premise of establishing a strong and efficient administration of British colonies in the region, especially in the face of Apartheid South Africa.

Brelsford (1960) argued that such a broad-based economic amalgamation was not only essential in economic terms, but would also form a foundation for an efficient public service which was needed in the formative days of colonial administration. This was seen as the most effective way of creating a multi-racial state which would counter the Nationalist Party's racial policies in South Africa (1948–1994). Indirectly, Apartheid South Africa was the architect of the Federation. The Federation was partially built on the background of fear and vulnerability of these three states in the face of radical white supremacy of the Nationalist Party (Brelsford, 1960). While this was the truth on paper, in reality, Southern Rhodesia emerged as the main beneficiary of this partnership since it was the administrative centre of the Federation. Investment into infrastructure, coupled with the existence of a high proportion of European settlers, strengthened Southern Rhodesia's economic power and hegemony of Northern Rhodesia and Nyasaland. It is therefore not surprising that a couple of hundred thousand Europeans – primarily in Southern Rhodesia – ruled over millions of black Africans through white minority rule. Hence the demise of the federation (in 1963) was largely instigated by a *horse-rider* type of relationship which emerged not only among partner countries of the federation but also between the colonial settlers and the indigenous black population.

What emerges clearly from this administrative amalgamation and economic platform in the form of the federation was the movement of labour which fed into the economic needs of the imperial economy. A number of regulations were enacted to support and stabilise labour needs – and some of them date back to the formative years of the Federation. What is interesting in these labour agreements was the gravitation towards the assimilation of immigrants by colonial governments irrespective of how they entered the country (see Table 4.3). However, it is important to note that these laws were only in the interests of the colonial governments because beyond labour stabilisation, immigrants were treated as subjects as opposed to the civil rights that were granted to white colonial settlers and immigrants. The societies so created remained polarised with most native people without the freedom of movement, let alone the right to vote. Part of the laxity in the immigration policies was born out of the refusal by immigrants to observe pass laws and national boundaries as seen in the enactment of the Lusaka Agreement of 1947. The job market represented the worst platform of exclusion where both immigrants and the host population were discriminated despite the fact that

Table 4.3 Labour Agreements

The Tripartite Labour Agreement of 1936 – Northern Rhodesia, Southern Rhodesia and Nyasaland
- Foreign workers were to be given two-year labour contracts.
- Employers viewed it as too short, limited workers' stay and affected their stability and efficient.

The Migrant Workers Act of 1948 – Northern Rhodesia, Southern Rhodesia and Nyasaland
- Migrants who had spent more than ten years of uninterrupted working period were supposed to remain permanently in that country.

The Lusaka Agreement of 1947 – Northern Rhodesia, Southern Rhodesia and Nyasaland
- Those who migrated illegally and worked over ten years without interruption to become bona fide residents of that country.

Source: National Archives of Zimbabwe, 1947.

they were the ones who were employing and selling their labour to generate capital that was invested into the luxury life of the settler population. Blake (1978:268) makes an interesting observation on this issue by noting that:

> In the first year of the federation, its GDP was an impressive £350 million; two years later it was nearly £450 million. Yet the average income of a European remained approximately ten times that of an African employed in the cash economy, representing only one third of local Africans.

While this was the same fate which the host members of society experienced, it represented a polarised society where immigrants were discriminated at two different fronts:

- Discrimination they suffered at the hands of the host society.
- Institutionalised discrimination at state level born out of racial and economic policies.

What emerges clearly from the Federation is the fact that labour movements were not limited to male labourers. Families were part of these movements, especially those who settled on commercial farms.

4.5 The SADC South African Development Community (SADC) States

The post-colonial period gave birth to the SADC. The Southern African Region is home to a number of bilateral agreements and regional bodies such as such as the Southern African Customs Union (SACU) and the South African Development Community (SADC). The SACU is an amalgamation of five countries (viz. Botswana, Namibia, Lesotho, Swaziland and South Africa) and it is the oldest customs union in the world. The SADC, which comprises 16 countries (see Table 4.4), is the biggest regional body in Southern Africa. In its formative year in 1980, the SADC comprised of Front-Line States (FLS) movement whose intention was to reduce the economic dependency on South Africa and provide coordinated support to countries that were still under minority rule. The founding member states of the FLS (Zambia, Tanzania, Swaziland, Mozambique, Lesotho, Botswana and Angola) later transformed this union into the SADC (in 1992) with the intention of pursuing development that would remove the vestiges of colonial rule and minority rule and reduce the dependency of member states on South Africa (Nshimbi et al. (2018). Soon after the collapse of the Apartheid system in 1994, South Africa became a member of this regional body.

The Southern African Region is not an ordinary region but an area endowed with vast mineral resources, agricultural land and manufacturing industries. Dominated by South Africa's economic power, the region has an aggregate per capita income of US$687 billion with an average per capita

Table 4.4 Countries in the SADC Region

Name of Country	Capital City	GDP (US)
Malawi	Lilongwe	7,663
Zambia	Lusaka	23,309
Comoros	Moron	119
Democratic Republic of Congo	Kinshasa	50,418
Tanzania	Dares alum	60,810
Angola	Luanda	89,603
Mozambique	Maputo	15,195
Namibia	Wind Hoek	12,541
Botswana	Gaborone	18,339
Lesotho	Maseru	2,289
Eswatini	Mbabane	4,471
Madagascar	Antananarivo	14,519
Seychelles	Victoria	1,580
Mauritius	Port Louis	14,048
South Africa	Pretoria	351,354
Zimbabwe	Harare	19,273

Source: Compiled by Author from various sources (2022).

income of US$1,940 per annum (Helen Suzman Foundation, 2019). Arrighi et al. (2010) summarise the socio-economic attributes of the region by emphasising that it is:

> Characterised by a combination of great mineral wealth, a white settler agriculture with no parallel elsewhere in Sub-Sahara Africa … which enabled the colonialists to create unlimited supplies of cheap permanent and temporary labour for mines and farms and, later, manufacturing industries in South Africa.

The colonial stage, in this context, was not only a platform for capital accumulation but also a stage for institutional development which manifested itself spatially through investment into infrastructure and development of towns and cities. Hence, the cities that emerged were not only cities for accumulation but also cities for consumption, where large numbers of immigrants were continually seeking economic asylum (among other determinants). Beyond colonial times, migration continues as people are pushed out of their homes due to conflict, natural disasters and the search for economic opportunities. Nshimbi and Fioramonti (2013) noted that between 1990 and 2010, most countries in the region have been increasingly receiving immigrants, with South Africa, Botswana and Namibia being the most favourite destinations, as shown in Table 4.5.

The SADC countries commit themselves to the African Union's (AU) effort to allow free movement of people on the continent. The initial Draft Protocol on Free Movement of Persons (of 1995) whose intention was to allow for visa-free entry rights into member countries for short visits, residence, establishment and work was rejected by some members states (such as

Table 4.5 Immigrants in the Region

Country	Estimated Migrants
South Africa	501,000 to 1.2 million
Botswana	10,000 to 76,000
Namibia	35,000 to 76,000

Author – adapted from Nshimbi and Fioramonti (2013).

South Africa, Botswana and Namibia). It was felt that the region was not yet ready for open intra-regional borders. The greatest fear of such a movement was the negative impact on the socio-economic infrastructure and the compromise on national immigration policies on receiving countries Nshimbi et al. (2018). At the moment, the region relies on the Facilitation Protocol of 2012 (ratified in 2015) which allows regional citizens visa-free entry for lawful purpose and visits of up to three months. Only six countries – South Africa, Botswana, Lesotho, Mozambique, Swaziland and Zambia – have ratified this protocol; hence, it remains unenforced due to lack of a two-thirds majority required to ratify it. This leaves the region to the regime of national immigration laws of individual SADC states as discussed in the chapters on case studies.

4.6 The City and Immigrants – Emerging Observations in the Region

As already alluded to in the preceding section, the overall objective of the study was to establish the extent to which cities of Southern Africa are inclusive, especially in the face of large numbers of immigrants being accommodated in member states. From the discussions presented so far in this chapter, two key strands concerning migrants and the city emerge – the city and the two-gate policy and the post-colonial protectionist city.

The City and the Two-Gate Policy: The African cities in the region (with the exception of those in Mozambique) were primarily products of British imperial interests which hinged on creating a supreme white bastion for their settlers. The presence of many white settlers in Zimbabwe and South Africa made this dream a reality. As such, the city was open to immigrants who met the strict civil requirements of the minority settlers – who were, in essence, whites. In Landau's (2011) own words, this represented the front gate which welcomed and assimilated immigrants who qualified into the governing minority. The back gate, on the other hand, was a 'deliberate institutionalised screen wall' which kept at bay unwanted immigrants while allowing access to cheap and docile temporary labour resident. For some reason, these segregation policies provided assimilation of immigrants within their respective exclusive zones. These cities had distinctive segregating features in their spatial layout which separated the white settlers from the indigenous people. As Christopher (1983) observes, the spatial expression of segregation was enforced through housing and labour laws that enabled access to other services. In the case of South Africa, the Native Urban Areas Act of 1923 (among other legislations and policies) entrenched the notion that the 'town was a European area in which there was no place for a redundant native' (Christopher, 1983, quoting Davenport, 1971). This saw the relegation of natives to Bantustans. Chipungu (2021), writing in the context of Harare, provides a vivid insight into these policies by noting that:

> The colonial government was not bent on creating an inclusive city with a racial balance in all social and economic spheres. For this reason, housing for Africans in the early years was based on short-term policies meant to accommodate immediate labour requirements only. Even when the need to stabilise labour became overwhelming in the late 1950s, housing was still provided on rental basis with a multiplicity of single dwellings dominating the housing production system and typology. More so, conventional housing production systems were popularly used in a bid to maintain the high standards of the European city.

This structured dual existence was spatially pronounced in countries like Zimbabwe and South Africa, whose high white settler population defended their supremacy even through protracted liberation struggles. Hence, it is not by accident that immigration during the colonial-apartheid period (i.e. after the Federation in Zimbabwe and during the Apartheid system in South Africa) remained minimal from within the region yet white settlers were encouraged.

The Post-colonial Protectionist City: On the other hand, those countries which were under the British Protectorate (such as Zambia, Malawi, Lesotho, Botswana and Swaziland) experienced limited racism – hence, the dominance of the indigenous population quickly gave way to self-rule. Colonial cities that emerged in these countries were equally spatially designed on racial lines but with a high presence of natives. It is these discriminatory policies driven by paternalistic reformism and racism that gave birth to African nationalism and region-wide resistance and armed struggles against colonial regimes. The fall of the Federation was followed by the independence of Northern Rhodesia and Nyasaland, while a full-fledged liberation struggle was wedged in Zimbabwe, Mozambique, Namibia and South Africa.

Box 4.2 Historical Insight into Immigration Dynamics and the City

- *Discrimination:* A pre-meditated discrimination system emerged out of the search for social, political and economic gains. This manifested itself in the form of the 'them and us', and was very pervasive between whites and blacks, but also between urban and rural. Indigenous Indians, blacks and immigrants occupied the lower echelons of the 'cast system'. But even among blacks, there was social friction, as seen in the use of derogatory names such as *makwerekwere* (foreigners), especially among the Shona (which reminds one as the continuation of the *ama-Hole* among the Ndebele), or the use of the term *kamu-Moscan* (derogatory term for Mozambican immigrants in Zimbabwe). Yet on the same plane, the Sotho and Tswana of Lesotho and Botswana were easily assimilated in South African towns. Legislations were enacted which legalised discrimination on the basis of race.

- *Exploitation:* At the centre of this historical platform was exploitation generated through low wages for labourers, structured wage systems and selective employment opportunities determined by skin colour (e.g. jobs for drivers and messengers were a preserve for Indians), while immigrants were always threatened with deportation and relegated to work as labourers on mines and farms.
- *Housing:* Housing was a multi-purpose 'tool' used to access the city and employment. For the minority settlers, their social status guaranteed them decent housing. Blacks and other immigrants were relegated to townships where tied-housing was provided. Hostels also emerged which were meant for male workers only, while those who could not secure any of the two ended up in squatter settlements where they endured continual harassment by law enforcement agencies.

Source: Authors (2022).

In summary, it can be argued that the narrative of immigration and its associated policies in Southern African cities is a narrative of labour movements and displacements born out of historical political and economic dynamics and consolidated by protectionist intentions. The liberation of some countries and the generation of mass movements of people displaced through wars in the region had implications for their recognition and stabilisation in receiving countries. Post-colonial Southern Africa is stuck between labour migrants seeking economic opportunities, on the one hand, and asylum seekers fleeing from civil wars, on the other. Hence what emerges are:

- Refugee camps which, in essence, are spatial enclaves which separate the host community from the immigrants. Indeed, this is justified and expected given the large numbers of people who are seeking refuge. The influx of any group of people comes with challenges pertaining to services, especially in urban areas where such services are expensive and already strained.

- Protectionist immigration policies that are bent on shielding the host communities. In the competitive modern economy, the interests of citizens take centre stage while those of immigrants are circumstantial, mainly ushered in through fears of social, economic and political security.

The historical trajectory of Southern Africa portrays a disjointed picture of immigrant inclusivity in cities and centres of production, largely driven by the selfish intentions of owners of means of production (especially mines and farms) and supported by political aspirants who were bent on building their careers backed by their colonial masters. In the ensuing process of social, political and economic determinants, the white minority across the whole breadth of the region emerged as 'trophies' of successful immigration, while the black majority were used as 'pawns' to fulfil these selfish motives. Temporary landmarks – which can be labelled 'black-dots' – emerged in cities in a bid to stabilise labour for the sole purpose of continuing the benefits from capital creation and accumulation that created a luxurious haven for the few. How did the post-colonial states respond to these acts of inhumanity, especially in cities and in the face of increasing immigrants? The various chapters on case studies respond to this question.

Note

1 South African History Online–SOHO@2021.

Bibliography

Arrighi, G., Aschoff, N. & Scully, B. 2010. Accumulation by Dispossession and its Limits: The Southern Africa Paradigm Revisited. *Studies In Comparative International Development*, 45, 410–438.

Blake, R. 1978. *A History of Rhodesia*. Eyre Methuen.

Brelsford, W. V. 1960. *Handbook to the Federation of Rhodesia and Nyasaland*. Cassell & Company, Ltd.

Chipungu, L. 2018. Migrant Labour and Social Construction of Citizenship in Lesotho and Swaziland. In Hangwelani Hope Magidimisha, et al. (Eds) *Crisis, Identity and Migration in Post-Colonial Southern Africa*. Springer.

Chipungu, L. & Magidimisha, H. H. 2020. *Housing in the Aftermath of the Fast Track Land Reform Programme in Zimbabwe*. Routledge.

Christopher, A. 1983. From Flint to Soweto: Reflections on the Colonial Origins of the Apartheid City. *Area*, 15, 145–149.

Crush, J. 2008. *South Africa: Policy in the Face of Xenophobia*. (MPI) Migration Policy Institute. Helen Suzman Foundation. South Africa.

Helen Suzman Foundation. 2019. South Africa.

Kowet, D. K. 1978. *Land, Labour Migration and Politics in Southern Africa: Botswana, Lesotho And Swaziland*. Nordiska Afrikainstitutet.

Landau, L. 2011. *Contemporary Migration to South Africa: A Regional Development Issue*. World Bank Publications.

Lesthaeghe, R. J. 1989. *Reproduction and Social Organization in Sub-Saharan Africa.* Univ of California Press.

Magidimisha, H. H. & Chipungu, L. 2019. *Spatial Planning in Service Delivery: Towards Distributive Justice in South Africa.* Springer.

Magubane, B. 1975. The Native Reserves. In Hi Safa, and Bm Du Toit (Eds) *Bantustans and the Role of the Migrant Labour System in the Political Economy of South Africa*, 1, 975.

Mitchell, J. C. 1985. Towards a Situational Sociology of Wage-Labour Circulation. In R.M. Prothero, and M. Chapman (Eds) *Circulation in Third World Countries*, 30–53, Routledge and Kegan Paul.

Nshimbi, C. C. & Fioramonti, L. 2013. A Region without Borders? Policy Frameworks for Regional Labour Migration Towards South Africa. *Nshimbi, Cc & Fioramonti, L. (2013) Miworc Report.*

Nshimbi, C. C., Moyo, I. & Gumbo, T. 2018. Between Neoliberal Orthodoxy and Securitisation: Prospects and Challenges for a Borderless Southern African Community. In H. H. Magidimisha, et al. (Eds) *Crisis, Identity and Migration in Post-Colonial Southern Africa*, 167–186, Springer International Publishing AG.

Palmer, R. 1979. *A History of Rhodesia.* JSTOR.

Trimikliniotis, N., Gordon, S., & Zondo, B. (2008). Globalisation and Migrant Labour in a 'Rainbow Nation': A fortress South Africa? *Third World Quarterly*, 29(7), 1323–1339.

Zulu, E. 2005. Interpreting Exodus from the Perspective of Ngoni Narratives Concerning Origins: Appropiating Exodus in Africa. *Scriptura: Journal for Contextual Hermeneutics in Southern Africa*, 90, 892–898.

5 Unpacking Migrant Laws in Gaborone

'The Gem of Africa'

5.1 Introduction

The town of Gaborone, formerly (until 1969) Gaberones, is the capital of Botswana. It was named after Kgosi (king) Gaborone (1820–1932), the chief of the Batlokwa people who had settled in the area known as Gaberones Village until 1965. The area was known in colonial times as Bechuanaland and came under British protection in March 1885 following pleas for assistance from Kgosi Khama III of the Bamangwato people following hostilities in the area between the Shona and the Ndebele tribes, which were further inflamed by the arrival of Boer settlers. The seat of the government was transferred there from Mafeking (now spelled Mafikeng), South Africa, in 1965, one year before Botswana became independent of Britain. Gaborone is located on the Cape-Zimbabwe railway and is the site of government offices, parliament buildings, health facilities, a thermal power station and an airport. It is the seat of the University of Botswana (founded 1976), and it also has a national museum and art gallery (1968), which includes the departments of natural history, archaeology and prehistory. Gaborone has a population of 231,592 according to 2011 data. Gaborone has a surface area of 169 square kilometres.

Today's Botswana was then administered as the northern Bechuanaland Protectorate and the Southern Crown colony of British Bechuanaland. On independence, British Bechuanaland was incorporated into Cape Colony in 1895 and is now part of South Africa whilst the Bechuanaland Protectorate became Botswana. It had the unenviable distinction of being the only territory in the world with its administrative centre lying outside its own boundaries, viz. at Mafeking in South Africa.

As independence loomed, it was clear that an administrative capital would be required within its own boundaries, and of the nine sites considered, the small settlement at Gaberones Village was deemed the most suitable as it already housed some government offices, had a railway line for transport, was close to Pretoria in South Africa, had a major source of water – the Ngotwane River – and, equally importantly, wasn't associated with any particular tribe. At that time, the then Gaberones Village had few other buildings save for the remnants of a colonial fort used as a base to police the area, a prison and a Government Rest House.

DOI: 10.4324/9781003184508-5

Once the location had been identified, work on the earth core fill Gaborone Dam began in 1963, securing a water source for the planned city, and the building of the city itself started in earnest in 1964, largely completed within two years in time to celebrate the birth of the new nation on 30 September 1966, when the town was renamed Gaborone. Today, Gaborone, with its population of just under a quarter of a million, boasts an international airport, stadium, large shopping malls and hotels together with government, industrial and financial hubs. It also serves as a gateway to rural Africa. It achieved city status in 1986.

Gaborone is controlled by the Gaborone City Council, the wealthiest council in Botswana (Nengwekhulu, 2008). Gaborone is the political centre of Botswana. Gaborone is one of the fastest growing cities in Africa. The growth of Gaborone, especially suburban growth, has caused much of the farmland surrounding the city to be absorbed into the city. Gaborone is the centre of the national economy. Several international companies have invested in the city. It has a comparatively lower unemployment rate. Gaborone is the fourth least expensive city for emigrants in Africa, following Addis Ababa, Ethiopia at 211th, Kampala, Uganda at 202nd and Windhoek, Namibia at 198th.

The city of Gaborone is home to over 10% of the population of Botswana (LeVert, 2007). The population growth rate of Gaborone is 3.4%, the highest in the country. This is most likely because the city has a more developed infrastructure, making it more liveable.[1] Seth (2008) states that Gaborone is one of the fastest-growing cities in the world. Much of the growth is based on net in migration from the rest of Botswana and neighbouring countries.

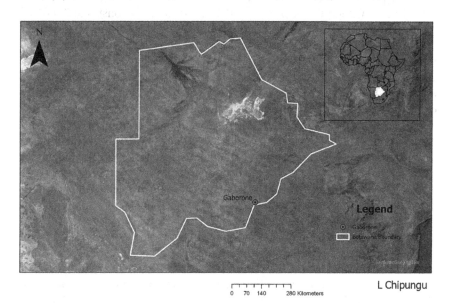

Figure 5.1 Map of Botswana Showing the Position of Gaborone

5.2 Migrant Profile in Gaborone

Gaborone is a destination for migrants from across the country and neighbouring states (Campbell and Crush, 2012), and given that it is located close to the border with South Africa, it is also a transit point for migrants from across the country and elsewhere in the Southern African Development Community (SADC) region. Not surprisingly, its growth rate of 3.4% is the highest in the country (CSO, 2005). The foregoing discussion implies that profiling the circumstances of migrants in Botswana and Gaborone in particular is not an easy undertaking considering that they constitute a heterogeneous group of both documented and undocumented, forced and voluntary migrants, short term (e.g. cross border traders) and long-term migrants (e.g. asylum seekers and refugees) (Ntseane and Mupedziswa, 2018).

Historically, Botswana played a significant role in hosting refugees from its neighbouring countries. As early as 1956, a significant influx of refugees came to Botswana fleeing from unfavourable socio-economic and political conditions in their countries (Parsons, 2008). By the early 1960s, another wave of hundreds of political refugees entered the country mostly from Namibia (Southall, 1984). Research shows that between 1967 and September 1969, thousands of Angolans fled their country and sought asylum in Botswana (Parsons, 2008). These were fleeing fighting between government and the rebel forces. By the time Botswana established Dukwi refugee camp in 1978, it is estimated that there were 20,000 refugees and asylum seekers residing at this camp (Ntseane & Mupedziswa, 2018).

Botswana hosts approximately 166,430 international migrants, of which the male population constitutes 91,038, while the female population constitutes 75,392. Children constitute 31.80% and the elderly constitute 2.00% of migrants. Around 1.71% of the migrants consist of refugee and asylum seekers (IOM, 2019). In 2004, Botswana is said to have been hosting over 300,000 mostly undocumented (forced) migrants (Lefko-Everett, 2004; Donnelly, 2004). However, the reaction of public officials to these migrants is that of implicitly denying their presence, excluding them from developmental plans, or tacitly condoning discrimination throughout the government bureaucracy and police. Despite being members of the community entitled to government resources (Götz, 2004), migrants are viewed and treated as an illegitimate drain on public resources. There is a distinct sense that current residents or 'ratepayers' deserve to be privileged over new arrivals or temporary residents. Some officials hold fast to the idea that migration worsens violent crime, disease and unemployment.

Displaced persons (asylum seekers and refugees) are prohibited to self-settle in urban areas. The majority of displaced persons or forced migrants are officially registered either as asylum seekers or as refugees and are either 'accommodated' at the Dukwi refugee camp (recognised refugees) or 'detained' at the temporary holding place, the Francistown Centre for Illegal Immigrants (FCII) (asylum seekers), where they receive basic provisions from the government of Botswana, UNHCR or other stakeholders. Botswana generally pursues a policy of restriction of movement of asylum seekers (i.e. detention) until granted status, and encampment of documented refugees, although on paper, some may be granted permission to settle outside the refugee camps.

Official records suggest that by 2011, the country was hosting 3,567 refugees from Algeria, Angola, Burundi, Congo, Eritrea, Ethiopia Kenya, Namibia, Rwanda, Somalia, South Africa, Sudan, Uganda and Zimbabwe. A majority came from Zimbabwe (1,007), Namibia (1,006), Somalia (555) and Angola (515) (Ntseane and Mupedziswa, 2018), and a most of these were stationed in the Dukwi refugee camp (UNHCR, 2014). Migrants who do not follow due processes and shunned official border entry points for clandestine crossing points (border jumping) into the country, often without proper documentation, often self-settle in urban areas like Gaborone. These migrants who spontaneously settled in the country or cities without permission or documentation are often called 'illegal immigrants or illegal foreigners'. In Gaborone, 'illegal immigrants' face a plethora of problems with housing, travelling, labour exploitation, access to health care, education and hostility from local communities.

Access to adequate housing is an uphill task for undocumented migrants in the city of Gaborone; hence, they tend to live in informal or low-cost settlements as tenants, where their landlords hardly ever ask them to produce immigration papers. They may rent a single room in high-density residential areas such as Naledi, often as a group, so they can share the burden of rentals. Locals at times exploit the situation and ask for exorbitant charges in rentals, knowing full well that these undocumented people are desperate. Consequently, many forced migrants struggle to meet their rental obligations. Not surprisingly, there is no question of undocumented migrants seeking official accommodation such as council (BHC) housing, the reason being that without proper immigration papers, the authorities would never entertain such a request. In these low-cost residential areas, overcrowding among forced migrants is an issue of major concern, with persons sharing a single room. Undoubtedly, sanitation becomes compromised in such circumstances, as many in these 'crowds' share facilities meant for much fewer people. Many have virtually no roof over their heads, and apparently some literary live on the streets of Gaborone, exposed to all the dangers associated with such an existence (Ntseane and Mupedziswa, 2018).

According to the interviews done with diplomats, a huge number of migrants come to Botswana in pursuit of economic opportunities, and this group of migrants include professionals, non-skilled and businesspersons. However, it has been gathered from the interviews that migrants face enormous challenges in accessing economic opportunities. The prohibition to work imposed on undocumented migrants in Botswana perhaps highlights the challenges migrants face to make ends meets. Regarding those migrants granted asylum, the new status on its own will not necessarily guarantee that they will secure employment. To get a job even at the cattle post, one needs a passport that can be used to apply for a work permit. Normally, job opportunities must be advertised in the press for a work permit to be processed. Yet, many of the migrants falling into this category do not have valid travel documents; hence, not surprisingly, they tend to lose out in this regard. The lack of employment opportunities exposes the undocumented migrants to poverty. In a study done by Moroka and Tshimanga (2009), forced and undocumented migrants in Botswana scored high on the 'living difficulties scale'. Consequently, they go to great lengths to raise money to help them secure basic needs. Ntseane and Mupedziswa (2018) corroborate that forced migrants in many situations have tended to engage in risky sex as a survival strategy or they engaged in transactional sexual relations. Thus, some engaged in sex for some sort of benefit, like free (or reduced rental) accommodation, being assisted to cross a border or some such favour. This of course renders them vulnerable to HIV infection and other STIs.

The finding from the interviews with diplomats reviews that labour exploitation of migrants, especially undocumented migrants, is rife in Botswana. Ditshwanelo (2005) observed that many refugees have only limited access to formal sector employment in Botswana. When employed, they are often paid minimal wages and are vulnerable to exploitation. Undocumented migrants based in Botswana are often content to do menial labour irrespective of their academic and professional qualifications and skills. As Lesetedi and Modie-Moroka (2007) have noted, while some undocumented migrants have marketable skills, they come to Botswana in the hope of getting decent jobs and end up being hired in jobs which are shunned by the Batswana. Many undocumented migrants are short-changed in the process by their local (informal) employers who tend to take advantage of their vulnerability. Some work but never get paid at the end, for their trouble. Undocumented migrants in Botswana, as those elsewhere, have to grapple with such challenges because they are a powerless lot as they lack social, political and economic 'clout' (Timngum, 2001). According to the US Department of State (2013), at the Dukwi refugee camp in Botswana, documented forced migrants are permitted to work outside the camp under certain circumstances. However, this contention contradicts that of UNHCR (2006), which observed that in Botswana there was little prospect of asylum seekers and refugees to be integrated locally, work or move freely in the country. The situation is of course even more daunting in the context of undocumented migrants. Without a work permit, their prospects for formal employment are virtually zero (Porta, 2012). The fact that undocumented migrants must pay for social services such as health care clearly makes their situation even more untenable. They desperately need access to employment, to be able to access the various social services. They are in a catch22 situation; without immigration documents; they cannot access decent jobs, and without decent jobs they cannot access social services. The way forward would be for them to regularise their stay but justifying their stay in Botswana would of course be a tall order, particularly given that many of them apparently fled economic rather than political upheavals in their home country.

Without an income, undocumented migrants cannot access the basic necessities of life. And yet, finding a job in Botswana is simply a tall order given that the country has an unemployment rate of over 17%. Migrants find it extremely difficult to secure a formal job in Gaborone. The law prohibits undocumented migrants from formal employment. The prohibition in Botswana appears to be essentially all encompassing, with the government policy requiring that asylum seekers be kept at the Francistown Centre for Illegal Immigrants (FCII) before being either transferred to the Dukwi camp (if their application for refugee status is successful) or deported (if their application is unsuccessful) (Lesetedi and Modie-Moroka, 2007). What is perhaps of some concern is that the conditions at the FCII reportedly leave nothing to be desired (Porta, 2012; Willet et al., 2015). Challenges at the detention centre have included overcrowding, mainly due to a high preponderance of over-stayers and immigrants living with mental disability (Masisi, 2009).

The interviews with diplomats reveal migrants' challenges in accessing basic health care in Botswana. Out of the various social services, perhaps the one with the greatest impact on the lives of forced migrants is health. In Botswana, the situation as regards documented (forced) migrants tends to be better than that of undocumented migrants. Undocumented (forced) migrants, however, tend to be (socially) excluded in regard to health care. For instance, to obtain medical attention from government health care institutions such as a hospital in Botswana, foreigners (whatever their status) pay slightly more than locals. And yet many of these foreigners, especially undocumented migrants, survive on odd jobs, usually those shunned by locals, or with no jobs at all. What this suggests is that many of them either forego treatment or delay as much as possible their visits to a health post for treatment because of the lack of money. They cannot visit a health facility without money for fear of being ridiculed by the health personnel. In some cases, by the time the migrants seek medical help, their condition would have deteriorated (Lesetedi and Modie-Moroka, 2007), often putting their lives in grave danger. Unlike the locals, undocumented migrants do not enjoy free access at government health institutions or to antiretroviral (ARVs) drugs, nor do they receive attention in the area of (free) PMTCT and related health issues. However, the US Embassy Botswana (2020)[2] states that in 2019 the government began allowing non-citizens, including refugees, to receive HIV and AIDS medication. The UNHCR facilitated refugee and asylum seekers' exit permit applications for medical referrals as necessary. Officials typically granted exit permits for three days; refugees found outside the camp without a permit were subject to arrest.

When it comes to education, as long as migrants remain undocumented, then access to education for their children is out of the question. Thus, children of undocumented migrants suffer for the sins of their parents in the sense that with their parents having no proper status documents, they (the children) are excluded from accessing education. The same applies to undocumented adults: if they are not able to produce documents attesting to their (legal) status in the country then they too can forget about enrolling at an adult education institution in the country. While no hard dates are available in this regard, it would probably be fair to surmise that there are potentially scores who fall into this category among the hundreds of undocumented migrants who have settled in the country over the years, some of whom may yearn to continue with their education. This clearly shows that access to education is a major challenge particularly for undocumented migrants in Botswana. Refugees in Botswana attend local primary and secondary schools alongside Batswana children. Many excel in their studies, but their options after secondary school are limited. The government offers higher education scholarships to cover all or part of tuition costs for Batswana students whose marks meet the required threshold, and some universities offer scholarships to students from lower income families. But refugees have traditionally not received these opportunities. Most come from families who cannot afford higher education, so they end up back in the Dukwi refugee camp, where almost all of the 1,010 refugees in Botswana live. There, they have few job opportunities.[3] In a very disturbing scenario on how the government of Botswana treats migrants, the US Embassy Botswana (2020) states that asylum seekers with children are transferred to the Dukwi refugee camp. International observers expressed concern that young children were sometimes separated from their parents while their cases were processed. In one case this included a family with eight minor children. International observers stated there was no access to education in the FCII, which, as of August, held 148 children younger than age 18.[4]

Considering the odds stacked against them, it is inconceivable how else the undocumented migrants in Botswana are able to sustain themselves. Due to limited resources, the government understandably cannot afford to provide migrants with free social services; hence, migrants in Botswana have virtually no access to social welfare services, including the safety nets rolled out by the government to mitigate the impact of poverty among vulnerable groups, including orphaned children, older people, people with disabilities, the indigent and the destitute, to ensure they lead relatively decent lives (Makhema, 2009; Ntseane and Solo, 2007). Consistent with the vision 2016 pillar of a caring nation, the government has over the years put in place a fairly robust social protection regime, which is the envy of many countries in the region (RHVP, 2011). Forced migrants, let alone those who are undocumented, virtually have no access to such programmes. Migrant children, no matter how desperate, cannot access such programmes as the orphans and vulnerable children (OVC) food basket, for example, nor can they access other destitute children welfare benefits. Benefits offered under the Destitute Policy, for example, enable both children and adult Batswana who are destitute to receive assistance such as food and school uniforms or qualify to be exempted from paying for certain services such as medical care. This facility is open only to citizens, and hence cannot be accessed by migrants. The government position is that due to limited resources, it cannot afford to extend this service to non-citizens, let alone undocumented ones. In some countries non-governmental organisations (NGOs) and faith-based organisations (FBOs) would be roped in. For Botswana, since the country attained middle-income status, the number of relief-oriented NGOs have been considerably reduced.

When it comes to co-existing with migrants, the local people have been doing so for a long time, especially with those they have employed. However, research has established that in Botswana xenophobic feelings, though subtle, do transcend local communities, cutting across gender, education, economic status and age (Morapedi, 2003), suggesting that these feelings may be deeply entrenched. It is noteworthy that some Batswana have been accused of a tendency to paint every foreigner with the same brush, irrespective of whether or not the individual possesses proper immigration papers and are in the country legally. This has at times promoted unbridled social exclusion. Interestingly, as is the norm across Africa, whites tend to be spared the hatred meted out against fellow black Africans. In various African countries, reasons for xenophobia have included dissatisfaction with life circumstances, fear of unemployment, insecurity about the future and low confidence in the way public authorities and political establishments work in each country (Shindondola, 2003).

Botswana was largely unprepared for the unprecedented volume of new immigrants, particularly the influx of Zimbabweans, in terms of physical infrastructure and border controls, as well as social absorption and integration strategies. The policy framework was at best inappropriate for the rate and the scale at which migration patterns are changing.

5.3 Migration Policies and the Regulatory Environment

Immigration to Botswana is regulated by the Immigration Act of 1966, which allows unrestricted entry for nationals from most countries in the region. In the early days of independence, Botswana operated an open-door policy which enabled scores of migrants to make their way into the country in large numbers (Campbell and Oucho, 2003). The economic growth and development came quickly to Botswana, and the country now boasts one of the strongest economies on the continent, political stability and great improvements in the quality of life of citizens. With this overwhelmingly positive outlook for the future, there was unprecedented influx of large numbers of migrants into the country, which forced the government to swiftly change its immigration policy, hence the introduction of tougher border controls and harsher punishment, particularly for illegal immigrants (Lefko-Everett, 2004), while levels of xenophobia are on the rise across the country, where this was once virtually unknown. With an increasingly educated and skilled domestic labour force, dependence on foreign skills is waning, and until recently, analysts might have predicted an eventual shift towards a more restrictive immigration policy. As a result, it now seems the 'Gem of Africa' is rapidly destined to be once again the exclusive property of the Batswana population. Like any government, issues of security and order are paramount for the government of Botswana. However, dealing with thousands of undocumented migrants has not been easy. The country resorted to a policy of arresting and deporting the 'undesirable elements'. To this end, apart from intensifying border patrols, combined patrols involving the police, the Departments of Immigration, Wildlife, the Customs and the Botswana Defence Force were also introduced (Lesetedi and Modie-Moroka, 2007). At one point, the government of Botswana apparently even embarked on a project to electrify selected (strategic) border areas that were being used mostly by Zimbabwe nationals to clandestinely enter the country. Occasional instances of stop-and-search by the police do occur, in efforts to flush out undocumented foreign nationals. There have been a few instances, too, of police raiding certain premises, including construction sites, in search of undocumented migrants. Those caught are detained and arrangements are made for their deportation in accordance with the law of the land.

Although Botswana has undoubtedly benefited from its open migration policy in the past, in the minds of many nationals, the consequences of openness today seem to outweigh its advantages. However, for the foreseeable future, Botswana is not facing the question of whether new migrants will arrive, but rather of how best to manage policing and control, protection, absorption and integration. This is to be done through the strengthening its policy, legislation and institutional framework.

Botswana became a state party to the 1951 Convention on the Status of Refugees[5] and its 1967 Protocol on 6 January 1969. Botswana became a signatory to the OAU Convention Governing Specific Aspects of Refugee Problems in Africa[6] on 16 May 1995. Unfortunately, no steps have been taken nationally to domesticate these treaties. Botswana is a dualist country and treaties have no force of law in Botswana until implementing legislation is promulgated.[7] Botswana's refugee law therefore remains archaic and anachronistic and much in need of review.

The Refugee Act, the Immigration Act,[8] the Constitution of Botswana, the Geneva Conventions Act[9] and other pieces of legislation affect the receipt and treatment of non-citizens in Botswana. When it comes to refugees, according to one diplomat (2021), asylum seekers and refugees are required by law to live in a refugee camp, as enshrined in the Refugee (Recognition and Control) Act of 1967.

- **The Refugee (Recognition and Control) Act 1967**
  According to the International Organization for Migration (2017), Botswana hosts approximately 2,845 refugees and asylum seekers. Botswana's policy towards refugees and asylum seekers is codified by the country's Refugee (Recognition and Control) Act (Refugee Act).[10] However, the law is now antiquated. The Refugee (Recognition and Control) Act of 1967 is control-oriented and not protection-oriented. Botswana's Refugee Act suffers from the problems plaguing similar control-oriented statues which were the norm in the region in the 1960s and 1970s. The first notable aspect of the above laws (control-oriented

statutes) is that that they were not comprehensive refugee legislation. Rather, they addressed selected aspects of the refugee problem. Second, the selected aspects did not so much relate to protection of refugees. Rather, as the long titles connote, they were mainly aimed at *controlling* refugees. The laws vest wide and discretionary powers to determine who is a refugee in the relevant Minister (Rutinwa, 2002). Botswana's control-oriented Act is devoid of many refugee rights contained in international treaties and conventions. Botswana's Refugees Act is silent on many rights, freedoms and duties of refugees.

According to the US Embassy Botswana (2020), in 2020, the government repatriated approximately 700 Zimbabwean refugees from the Dukwi refugee camp, using a combination of voluntary removals and deportation. Many of the Zimbabweans had been living at the camp for more than a decade. Following UNHCR intervention into the fast-tracked refugee status determination process for Zimbabweans, the government agreed to a more thorough process. The government determined, with the UNHCR's agreement, that ten persons warranted further protection due to lesbian, gay, bisexual, transgender, queer or intersex (LGBTQI+) status or because they were senior-ranking members of Zimbabwean political opposition groups. Most of the remaining 171, who were deemed no longer needing international protection, were left voluntarily, although the government ultimately deported 36 persons in April, including military deserters who expressed fear of retribution upon return.[11]

- **Botswana's Constitution of 1966 with Amendments through 2016**

Though the constitution does not directly make provision for immigrants, it provides a fundamental basis for protection of persons in Botswana. That could possibly include immigrants. Chapter 2 of the Constitution provides for the Protection of Fundamental Rights and Freedoms of the Individual (ss 3 – 19). Section 3 says:

> Whereas every person in Botswana is entitled to the fundamental rights and freedoms of the individual, that is to say, the right, whatever his or her race, place of origin, political opinions, colour, creed or sex, but subject to respect for the rights and freedoms of others and for the public interest to each and all of the following, namely—
>
> a   life, liberty, security of the person and the protection of the law;
> b   freedom of conscience, of expression and of assembly and association; and
> c   protection for the privacy of his or her home and other property and from deprivation of property without compensation, the provisions of this Chapter shall have effect for the purpose of affording protection to those rights and freedoms subject to such limitations of that protection as are contained in those provisions, being limitations designed to ensure that the enjoyment

of the said rights and freedoms by any individual does not prejudice the rights and freedoms of others or the public interest.[12]

Section 15 of Chapter 2 provides for protection from discrimination on the grounds of race, etc. It says:

1. Subject to the provisions of subsections (4), (5) and (7) of this section, no law shall make any provision that is discriminatory either of itself or in its effect.
2. Subject to the provisions of subsections (6), (7) and (8) of this section, no person shall be treated in a discriminatory manner by any person acting by virtue of any written law or in the performance of the functions of any public office or any public authority.
3. In this section, the expression 'discriminatory' means affording different treatment to different persons, attributable wholly or mainly to their respective descriptions by race, tribe, place of origin, political opinions, colour, creed or sex whereby persons of one such description are subjected to disabilities or restrictions to which persons of another such description are not made subject or are accorded privileges or advantages which are not accorded to persons of another such description.
4. Subsection (1) of this section shall not apply to any law so far as that law makes provision—
    a. For the appropriation of public revenues or other public funds.
    b. With respect to persons who are not citizens of Botswana.
    c. With respect to adoption, marriage, divorce, burial, devolution of property on death or other matters of personal law.
    d. For the application in the case of members of a particular race, community, or tribe of customary law with respect to any matter whether to the exclusion of any law in respect to that matter which is applicable in the case of other persons or not; or
    e. Whereby persons of any such description as is mentioned in subsection (3) of this section may be subjected to any disability or restriction or may be accorded any privilege or advantage which, having regard to its nature and to special circumstances pertaining to those persons or to persons of any other such description, is reasonably justifiable in a democratic society.
5. Nothing contained in any law shall be held to be inconsistent with or in contravention of subsection (1) of this section to the extent that it makes reasonable provision with respect to qualifications for service as a public officer or as a member of a

disciplined force or for the service of a local government authority or a body corporate established directly by any law.
6 Subsection (2) of this section shall not apply to anything which is expressly or by necessary implication authorized to be done by any such provision of law as is referred to in subsection (4) or (5) of this section.
7 Nothing contained in or done under the authority of any law shall be held to be inconsistent with or in contravention of this section to the extent that the law in question makes provision whereby persons of any such description as is mentioned in subsection (3) of this section may be subjected to any restriction on the rights and freedoms guaranteed by sections 9, 11, 12, 13 and 14 of this Constitution, being such a restriction as is authorized by section 9(2), 11(5), 12(2) 13(2), or 14(3), as the case may be.
8 Nothing in subsection (2) of this section shall affect any discretion relating to the institution, conduct or discontinuance of civil or criminal proceedings in any court that is vested in any person by or under this Constitution or any other law.
9 Nothing contained in or done under the authority of any law shall be held to be inconsistent with the provisions of this section—

   a if that law was in force immediately before the coming into operation of this Constitution and has continued in force at all times since the coming into operation of this Constitution; or
   b to the extent that the law repeals and re-enacts any provision which has been contained in any written law at all times since immediately before the coming into operation of this Constitution.[13]

- **The Geneva Convention Act**
  Act no. 28 of 1970 enables effect to be given in Botswana to certain international conventions made at Geneva on 12 August 1949. Under the act, no contracting state shall expel or return a refugee in any manner whatsoever to the frontiers of territories where his life or freedom would be threatened on account of his race, religion, nationality, membership of a particular social group or political opinion.

5.4 Migration Institutions and Governance

The key players in terms of welfare provision are the Government of Botswana, the UNHCR and the Botswana Red Cross Society.

- **United Nations High Commission for Refugees (UNHCR)**
  The UNHCR encourages the participation of its officials in status determination. The rationale for this is the need to allow the UNHCR to monitor closely matters of status, the entry and removal of refugees and

participation in the identification of those who should benefit from refugee status (Goodwin-Gill and McAdam, 2007) and the provision of up-to-date information regarding the general situation in an applicant's country of origin.[14]

Asylum seekers should not be denied the opportunity to communicate with the UNHCR.[15] The UNHCR should also be afforded the right to present its view in the exercise of its supervisory responsibility to any competent authority regarding individual applications for asylum at any stage of the procedure under Article 35 of the 1951 Convention.[16]

Although the Refugee (Recognition and Control) Act of 1967 is silent on this, the practice on the ground is that applicants for refugee status received by the UNHCR are provided with an initial interview with a protection officer. The UNHCR also has *ad hoc* representation in the Refugee Advisory Council. The role of the UNHCR in the Refugee Advisory Council is to sit as an *ex officio* member and advise on refugee law and provide up-to-date information in so far as is possible on an applicant's country of origin. The UNHCR reports that this partnership with the Refugee Advisory Council works remarkably well. It is submitted that the Act should be revised to formalise participation by the UNHCR in the status determination procedure.

- **Botswana Red Cross Society (BRCS).**

The BRCS is not an NGO. It is a National Society (NS) that was constituted on the basis of international and domestic laws by virtue of the government being signatory to the Geneva Conventions. The mandate of the BRCS as stipulated in the Act is to complement government work on health, social services and disaster management. The BRCS forms part of the International Federation of Red Cross and Red Crescent Societies (IFRC), which is the largest humanitarian network operating in 190 countries.

The IFRC and the International Committee of the Red Cross (ICRC) make up the Red/Crescent Societies. The National Societies are found in 190 countries and have been accorded auxiliary status to their state parties (governments) in the humanitarian field. State parties are signatory to the Geneva Conventions and additional protocols.

The National Societies carry out their activities in conformity with their own statutes and national legislation in pursuance of the mission of the statutes of the International Red Cross and Red Crescent Movement and in accordance with the Fundamental Principles.

The National Societies support the public authorities in their humanitarian tasks according to the needs of the people of their respective countries. The National Societies are viewed as reliable partners to their state parties – partners who will provide service based on its unique capacity to rapidly mobilise considerable human and material resources, including at the community level and comprising volunteers.

As members of the International Conference of Red/Crescent Movement, governments participate directly in the framing of movement policies and legal frameworks. This gives the Red Cross/Red Crescent predictability and transparency towards the states which may not always exist with other organisations. The 28th International Conference accepted the concept of a 'balanced relationship' between governments and National Societies.

- **Ministry of Nationality, Immigration and Gender Affairs**
  The department's responsibilities include facilitating the movement of people, issuance of passports, visas and residence permits and border control and management.

5.5 Summary

What can be deduced from the narrative on Botswana is that like any other country in the region, it has a proper institutional framework for migration. The institutional framework is informed by the laws of the country, regional protocols and the international laws. Various governmental departments and non-governmental organisations are involved in migrations issues. While the country is open to immigrants as per its migration policies, it has very strict laws when it comes to undocumented immigrants. While some argue that its migration policies need to change in line with global dynamics, the influx of illegal immigrants especially from within the region remains a big problem. Its stable economy, which makes the country to be among the best in the region, will always make it one of the preferred destinations to immigrants from within the region and abroad.

Notes

1 Central Statistics Office (February 2005). '2001 Population Census Atlas: Botswana'. Gaborone, Botswana.
2 https://bw.usembassy.gov/human-rights-report-on-botswana/
3 https://reliefweb.int/report/botswana/hope-and-opportunity-refugee-students-botswana
4 https://bw.usembassy.gov/human-rights-report-on-botswana/
5 General Assembly resolution 429 (V) (1951).
6 UN Treaty Series No. 14691 (1969).
7 7 *See Good v The Attorney General* (2005) (1) BLR 462 (HC); *Attorney General v Dow* (1992) BLR 119 at 152; *Ramantele v Mmusi* (2013) CACGB 104–12 (unreported) at para 69; and BOPEU *and Ors v Minister of Home Affairs* MAHLB 674-11 at para. 190–209.
8 Cap 25:02 Laws of Botswana.
9 Act No. 28 of 1970.
10 Cap 25:03 Laws of Botswana.
11 https://bw.usembassy.gov/human-rights-report-on-botswana/
12 https://www.constituteproject.org/constitution/Botswana_2016.pdf?lang=en
13 https://www.constituteproject.org/constitution/Botswana_2016.pdf?lang=en
14 Art 8(2)(b) 1951 Convention.
15 Art 10(1)(c) 1951 Convention.
16 Art 21(1)(c) 1951 Convention.

Bibliography

Campbell, E. & Crush, J. 2012. Unfriendly Neighbours: Contemporary Migration from Zimbabwe to Botswana.
Campbell, E. K. & Oucho, J. O. 2003. *Changing Attitudes to Immigration and Refugee Policy in Botswana*. Population Studies and Research Institute, University of Nairobi.
CSO 2005. *2001 Population Census Atlas: Botswana*. Gaborone: Centrals Statistics Office.
Ditshwanelo 2005. Rights of Minority Groups. Botswana Centre For Human Rights, Working Paper. Ditshwanelo.
Donnelly, J. 2004. Zimbabwe Woes Spill Across Border. *Zw News* (2 March 2004), 2.
Goodwin-Gill, G. S. & McAdam, J. 2007. *The Refugee In International Law, 3*. Baskı, New York. Oxford Yayınevi.
Götz, G. 2004. The Role of Local Government Towards Forced Migrants. In *Forced Migrants in the New Johannesburg: Towards a Local Government Response*. Johannesburg: Forced Migration Studies Programme.
IOM. 2019. *IOM Regional Strategy for Southern Africa 2019–2023*. International Organization of Migration.
Lefko-Everett, K. 2004. Botswana's Changing Migration Patterns. https://www.migrationpolicy.org/article/botswanas-changing-migration-patterns
Lesetedi, G. N. & Modie-Moroka, T. 2007. Reverse Xenophobia: Immigrants Attitudes towards Citizens in Botswana. In *African Migrations Workshop: Understanding Migration Dynamics in the Continent*. Ghana: Centre for Migration Studies, University of Ghana, Legon-Accra.
Levert, S. 2007. *Botswana*. Marshall Cavendish.
Makhema, M. 2009. Social Protection for Refugees and Asylum Seekers in the Southern Africa Development Community (SADC). *The World Bank, Social Protection and Labour Discussion Paper*, 15, 3–39.
Masisi, P. 2009. Overstaying Immigrants' Burden Center. *Daily News*, 17.
Morapedi, W. 2003. *Post Liberation Xenophobia and the African Renaissance: The Case of Undocumented Zimbabwean Immigrants into Botswana C 1995–2003*. Gaborone: History Departmental Seminar, In University of Botswana
Moroka, T. & Tshimanga, M. 2009. *Barriers to and Use of Health Care Services among Cross-Border Migrants in Botswana: Implications for Public Health*. International Journal of Migration, Health and Social Care, 5, 33–42.
Nengwekhulu, R. 2008. *An Evaluation of the Nature and Role of Local Government in Post Colonial Botswana*. University of Pretoria.
Ntseane, D. & Mupedziswa, R. 2018. Fifty Years of Democracy: Botswana's Experience in Caring for Refugees and Displaced Persons. *International Journal of Development And Sustainability*, 7, 1408–1427.
Ntseane, D. & Solo, K. 2007. Access to Social Services for Non-Citizens and the Portability of Social Benefits Within the Southern African Development Community—Botswana Country Report. Background Paper for Joint IDS/World Bank Research Project.
Parsons, N. 2008. The Pipeline: Botswana's Reception of Refugees, 1956–68. *Social Dynamics*, 34, 17–32.
Porta, J. 2012. *Just We Look for Peaceful Life: Forced Migrants in Botswana and Decision Making in the Transit Experience*. University of Toronto Scarborough.
RHVP (Regional Hunger and Vulnerability Programme) 2011. *Social Protection in Botswana – A Model for Africa?* Johannesburg: RHVP.

Rutinwa, B. 2002. Asylum and Refugee Policies in Southern Africa: A Historical Perspective. In *Legal Resources Foundation, A Reference Guide to Refugee Law and Issues in Southern Africa*, 1–10, Southern African Regional Poverty Network (SARPN).

Seth, W. 2008. Major Urban Centres. In Seth, W. (Ed.). *Botswana and its People*, 44–46. South Africa: New Press.

Shindondola, H. 2003. *Xenophobia in South Africa and Beyond: Some Literature for a Doctoral Research Proposal*. Johannesburg: Rand Afrikaans University.

Southall, R. 1984. Botswana as a Host Country for Refugees. *Journal of Commonwealth & Comparative Politics*, 22, 151–179.

Timngum, D. A. 2001. *Socio-Economic Experiences of Francophone and Anglophone Refugees in South Africa: A Case Study of Cameroonian Urban Refugees in Johannesburg*. South Africa: University of the Witwatersrand.

UNHCR. 2006. *Country Operations Plan (Botswana)*. Gaborone.

UNHCR. 2014. *Unhcr Global Report – Southern Africa Sub Regional Overview*. Geneva.

US Department of State. 2013. *Human Rights Report, Botswana*. Bureau of Democracy, Human Rights and Labour.

Willet, S. M., Schmid, J. & Hudson, D. J. 2015. *Voices of Kagisong: History of a Refugee Programme in Botswana*. Gaborone: Bay Publishing.

6 Malawi
A Retreat into Lilongwe the 'City Centre'

6.1 Introduction

Lilongwe city has been the capital of Malawi since 1975. Lilongwe, which was named after a nearby river, is located on the inland plains and is one of the largest cities in the country. It became a colonial district headquarters in 1904. The emergence in the 1920s of a major tobacco industry in the surrounding areas, along with its location at the junction of several major roadways, increased Lilongwe's importance as an agricultural market centre for the fertile Central Region Plateau. In 1965 Malawi's first president, Hastings Kamuzu Banda, selected it as an economic growth point for northern and central Malawi. Lilongwe is located in the centre of this long, narrow nation, making it easier for the country's citizens to reach it as opposed to the previous capital in the far south section of the nation. Lilongwe is also close to the countries of Mozambique and Zambia.

The city started life as a small village on the banks of the Lilongwe River, and became a British colonial administrative centre at the beginning of the 20th century. Due to its location on the main north-south route through the country and the road to Northern Rhodesia (now Zambia), Lilongwe became the second largest city in Malawi. In 1974, the capital of the country was formally moved from Zomba (the third largest city today in Malawi) to Lilongwe. Although Lilongwe is the official capital of Malawi and has grown immensely since 1974, most commercial activity takes place in Malawi's largest city, Blantyre. Recently, as part of political restructuring, the parliament has been shifted to Lilongwe and all parliament members are required to spend time in the new capital. Lilongwe is now the political centre of Malawi, but Blantyre remains the economic capital.

Since becoming the capital of Malawi, Lilongwe has grown dramatically, partly because of the numerous Europeans and South Africans residing in the city. A number of non-governmental organisations (NGOs) in Southern Africa are located there. The city is also the headquarters for British and United States foreign aid workers, diplomats from dozens of nations and international corporations. Although Lilongwe is the political centre of the country, Blantyre in the southern part of the country remains the economic and commercial centre of Malawi.

DOI: 10.4324/9781003184508-6

Lilongwe was not the first capital of Malawi; that title belongs to Zomba. For centuries, Lilongwe was a small fishing village. In the early 20th Century when British Colonial officials took control over and named the new colony, Nyasaland, they declared Zomba to be the administrative centre. In 1964 the country's name was changed from Nyasaland to Malawi when it gained its independence and Zomba remained the capital. Long-time Malawi President Hastings Banda, who ruled the nation from 1964 to 1994, moved the capital to Lilongwe in 1975 partly because of its central location but also because he was born just north of the city.

The city is divided into two sections, Old and New Town. Old Town is the area where the original fishing village was located and today it continues to have the appearance of a traditional African settlement with cafes, open air markets and numerous small shops and other businesses. New Town came into existence after Lilongwe became the capital and has modern buildings, including the national legislative building, government ministries, embassies, hotels, banks and office towers. The city centre has a nature sanctuary that cares for rescued and injured wild animals.

The official languages of Lilongwe and the nation of Malawi are English and Chichewa. The country has two major religions, Christianity – with most Christians being Roman Catholic or Presbyterian – and Islam, with most Muslims belonging to the Sunni sect. Although Lilongwe is the capital, many of its citizens are poor. In 2012 Lilongwe city had an estimated population of 781,000 people, making it the largest city in the country.

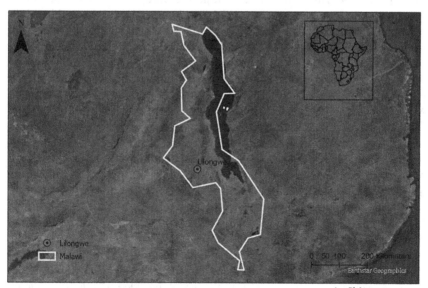

Figure 6.1 Map of Malawi Showing the Position of Lilongwe

As compared to other African capitals, Lilongwe is politically stable, safe and quiet. Many European and South African expatriates live in Lilongwe, and many NGOs (Care International, Plan International, Concern, UNHCR, UNFAO, WFP, Population Services International, the UNC Project, World Camp), international aid organisations (Peace Corps, USAID, DFID) and international corporations, particularly tobacco-related firms, operate out of Lilongwe. As a result, most Western visitors will find the city to be accommodating and friendly.

In Lilongwe, as opposed to rural Malawi, one can live, work or vacation in a manner that most Westerners would consider typical, if not luxurious. However, most of Lilongwe's Malawian citizens live on just a few dollars a day and many are unemployed. The population of Lilongwe has grown as villagers, including young orphaned children, from the surrounding rural areas have relocated to the capital in search of often non-existent jobs and the unattainable quality of life enjoyed by government officials, NGO and other international workers, and expatriates. Despite the highly visible class differences, most of the city's residents go about their lives in relative harmony. Street crime is uncommon, but begging and street hustling are not.

6.2 Migrant Profile in Lilongwe

Migration dynamics in Malawi are complex as it is a country of origin, transit and destination for thousands of migrants. The country is particularly affected by mixed migration flows and related complex smuggling routes from the Horn of Africa to Southern Africa. Malawi was host to tens of thousands of Mozambican and Rhodesian refugees in the 1970s and 1980s, but most of these were repatriated after independence was achieved in Mozambique and Zimbabwe. Migration into Malawi has noticeably increased recently, with the highest number of documented immigrants coming from Pakistan and India. Most of the Asian immigrants are either in business or work permits. A significant number of Asian immigrants who have received permanent resident or naturalisation permits have been legally residing in Malawi for extended periods, either in business or work permits. Immigrants from the Great Lakes countries find pathways to permanent residence and work permits from initial status as refugees and asylum seekers. Other leading neighbouring countries contributing significant numbers of migrants are Mozambique, the United Republic of Tanzania, Zambia and Zimbabwe. These countries, with the exception of Zimbabwe, all border Malawi and have populations that share cultural heritage along the borders. They also have vibrant economic activities across their borders with Malawi. Therefore, migration from these countries tends to be family and trade related.

In the past decades, Malawi has also seen an increased number of immigrants largely from the Horn of Africa and the Great Lakes region (Kainja, 2012), Mozambique, Somalia, Ethiopia, Tanzania, Democratic Republic of Congo (DRC), Burundi, Rwanda and Sudan (Nkhoma, 2014). Migrants from these countries tend to seek refuge and asylum-seeker status and are therefore mostly either at the Karonga reception centre in the north or at the Dzaleka refugee camp which is located in the Dowa District near Malawi's capital city Lilongwe and hosts around 44,000 refugees and asylum seekers. More than 3,000 Mozambican asylum seekers are in the Luwani refugee camp, in the south. The UN estimates there are around 2,000 refugees residing outside the camp at Dzaleka. The number of people who have fled to Malawi has risen from almost 17,000 in 2013 to more than 37,000 in March 2018 and new asylum seekers, particularly from the Democratic Republic of the Congo, are arriving each month. However, official refugee statistics are debatable with various institutions coming up with different figures. Until the late 1990s, Mozambique was the largest generator of immigrants who relocated themselves in different countries within the region as refugees (Callamard, 1994). While the latter come both as regular and illegal immigrants to exploit business opportunities in the region, the former are more or less refugees. While most of the immigrants are legally located in the country, there has been an increased number of immigrants who locate themselves illegally without due documents. As of 2010, the total number of illegal immigrants in the country was reported to be 279,000 (Nkhoma, 2014). The report indicated that the country was receiving an average of 5,000 illegal immigrants per annum, the majority of which entered the country through the northern districts of Karonga, Chitipa, Nkhata Bay, Rumphi, Nkhotakota and Salima. Between 2008 and 2009 over 350 illegal immigrants were arrested and 2,000 were deported to their original homes. Despite the arrests and deportation, the number of illegal immigrants continued to increase in the country. For example, over 500 illegal immigrants were convicted and 3,000 deported between 2009 and 2010 (Nkhoma, 2014).

According to the UN (2017),[1] Malawi hosts approximately 237,104 international migrants, of which the male population constitutes 112,936, while the female population constitutes 124,168. Children constitute 50.50% and the elderly constitute 2.90% of migrants. Around 10.84% of migrants consist of refugee and asylum seekers. By mid-year 2020, the international migration stock for Malawi stood at 1% of the whole population (191,400), while the total number of emigrants stood at 311,100, culminating in a net migration of –80.3 thousand. The percentage of female migrants in the international stock was 51.1%, indicating an increase in the feminisation of migration in Malawi (IOM, 2021; Refugee and Migrants 2020).

The majority of youth migrants, especially those of refugee status, have spent all their lives in Malawi, taught the same school curriculum as the local population, surrounded by a local culture and among a local people yet not free to integrate as local citizens. Without the inherent rights and freedoms of citizens, the younger generation of refugees is more and more despondent. As a signatory to the 1951 Refugee Convention, Malawi is obliged to adhere to the Convention but, as was its right, made nine reservations. The reservations pertain to the provisions of wage-earning employment, public education, labour legislation, social security and freedom of movement for refugees within Malawi. These reservations pose complex challenges, especially for adolescents entering into adulthood who wish to seek higher education, gain employment, marry and begin families. A sentiment shared by the young adults in Dzaleka is that the current situation and the challenges they face entering into adulthood are largely out of their control.

Malawi has the highest increase in urbanisation rate in the world at about 6% per year (Kloukinas et al., 2020). Lilongwe's spectacular growth rates are, however, only partly based on rural–urban migration flows. As Potts has pointed out, much of Lilongwe's exorbitant growth rates in the 1970s were due to urban–rural boundary changes in the course of its establishment, during which large rural areas were assigned for the further extension of the city (Potts, 1986). The growth rates also have to be dealt with cautiously when compared to the overall migration behaviour in Malawi. Migration patterns are very volatile and mainly circulatory, which in turn would suggest volatile, fluctuating urbanisation. This is to a certain extent confirmed by Potts in her study of Lilongwe in the 1980s. Whereas gross rural–urban migration was high, most of the migration movements into town and within rural areas were short term, i.e. seasonal or temporary, rather than for lifetimes or for longer periods (Potts, 1986). However, this increase in urbanisation undoubtedly increases the demand for housing as more people are coming to the city.

In urban areas, only 20% of the houses are delivered through the formal sector (Manda, 2013). Kloukinas et al. (2020) state that the majority of the households have to cover their housing needs by their own means and with limited access to loans and micro-financial tools. In fact, individual builders dominate housing construction, with 90% of houses self-built (Rust and Gavera, 2013). Most people rent a house or own one in the informal sector, as it can provide houses about 60 times cheaper than the formal sector (Rust and Gavera, 2013). Kloukinas et al. (2020) state that informal construction accounts for more than 90% of housing in the country, as access to formal housing is too expensive for most of the population. In Lilongwe, the lack of affordable housing forces low-income groups to erect poor houses in informal settlements. The majority of the population opt for the informal sector because of the urgent need for housing and the affordability, which surpasses the long-term security advantages (UN-Habitat, 2010). By default, the majority of migrants cannot afford formal housing; hence, they also turn to informal housing. Whilst Lilongwe is host to a long-term refugee community, the city itself is populated by many low-income individuals, with 76% of its inhabitants living in informal settlements, meaning their housing is often low quality and insecure. This situation has implications for a whole range of other areas of the community.

The socio-economic rights in Malawi are written into the constitution. The constitution ensures a small number of socio-economic rights, i.e. the right to education, economic activity and labour. Although school fees were abolished in 1994, over 10% of children of primary school age do not attend school and only 40% of the pupils enrolled in Standard 1 reach Standard 4. The quality of education is limited by the size of classes (107 pupils to one teacher) and the lack of facilities and materials. With regard to tertiary education, migrants, including refugees, have to pay the full fees at university, while the government subsidises the education of national students. Studies about the education situation in Malawi indicate that the refugees and other migrants generally experience worse conditions than the citizens. The situation is also dependent on whether the migrants live inside or outside the camp. Access to education by refugees and asylum seekers is limited. Both in the camp and in urban areas, they have access to pre-school, primary and secondary school, but are not allowed to pursue tertiary education at the nation's university. However, citizens also struggle to access the university. Malawians, refugees and asylum seekers go to the same public schools. There is a generally held view that the quality of state education in Malawi is not that high, yet private school fees are exorbitantly expensive for the refugees and other poor migrants to afford.

Refugees are not allowed to attend public schools outside the camps at primary level. The government provides schools for refugees within the camps. On the other hand, children from the local communities (citizens) close to the camps are allowed to attend primary school within the camps. In addition to basic education, refugees in the camp are also provided with vocational skills training, including income-generating activities. The respondents mentioned that the training courses offered at the camp include tailoring, computer courses, literacy classes, carpentry, Swahili and WUSC language classes.

Despite the constitution of Malawi not guaranteeing access to health care, Malawi has committed to allocating a significant portion of government expenditure to health. Over the past few years, the Malawian government has made efforts to improve the quality of the health service, increasing the annual budgetary allocation to improve the availability of drugs as well as infrastructure development, which includes various new clinics and hospitals, or hospital extensions (such as maternity wings). However, the delivery of services continues to be hampered by a persistent lack of sufficient drugs in the clinics and hospitals. The issue of financial constraints hits the refugees hard. Most poor migrants do not have any alternative but to visit public clinics or hospitals because of the financial implications of going to private facilities. Insufficient staffing is a major challenge in most public hospitals. The camp's hospital ratio is 200 patients per clinician on a busy day. This affects the quality of services offered. Issues to do with discrimination are also of major concern. There are reports of discrimination based on racism, nationality, tribalism and language by certain local medical staff. This is manifested in the languages used, case handling and general attitude towards the sick. Every refugee is referred to as 'a Burundi'. The UNHCR does not provide medical treatment for refugees who live outside the camp.

Sustained regional instability means that Malawi will continue receiving refugees and asylum seekers and also be used by organised syndicates as a staging ground for smuggling and possible human trafficking. Evidence has already shown that the country is used as a 'transit' ground for irregular migrants destined for other Southern African countries (Ndegwa, 2015). As of May 2021, there were 52,258 refugees and asylum seekers in Malawi,[2] coming from the Democratic Republic of the Congo (31,551), Burundi (11,241), Rwanda (6,939), Mozambique (4) and 2,523 more from other nationalities. Malawi has an encampment policy, and all refugees and asylum seekers are hosted in the only existing camp, the Dzaleka refugee camp. This camp, initially designed for 10,000 people, now has more than 48,000 people, posing serious health risks. Some of the challenges refugees face in the Dzaleka refugee camp are lack of freedom of movement, poor sanitation and congestion. Although refugees experience some obstacles, the Government of Malawi grants voting rights to foreigners, including refugees after seven years of residence. According to the 1989 Refugee Act, refugees are required to stay in camps where their needs are attended to with assistance from the UNHCR. Refugees are provided with basic assistance such as shelter, food, water, relief items, education and health care, but they do not enjoy the right to work and move outside of camps. However, this does not mean other migrants do not face challenges in accessing services and opportunities to be able to integrate well in the country, especially in urban area like Lilongwe. Similar to the local people, migrants face challenges in accessing adequate housing, education, health care and economic opportunities. This also affect their integration process.

Although historically common, international migration from Malawi decreased in the 1960s after independence when President Banda discouraged the migration of workers, especially to South Africa, migration to South Africa did increase again in the 1990s, mainly through informal means (Beegle and Poulin, 2013). It is also important to highlight that Malawi is more of a migrant sending country than it is a receiving country. In 1990, half of all Malawian emigrants lived in Zimbabwe, followed by Zambia (14%), and South Africa (11%). However, because of the subsequent adverse economic situation in Zimbabwe and the end of hostilities in Mozambique, by 2015 the share of Malawian immigrants in Mozambique rose from 2% to 25% (77,488). In the same year, the share of Malawians living in other African countries was as follows: Zimbabwe 102,849; South Africa, 76,605; Zambia, 11,258; Tanzania, 6,907; and Botswana, 4,596. Outside Africa, Europe hosts about 19,557 Malawians, mostly in the UK (17,871), while Canada and Australia are home to 981 and 1,266 Malawians, respectively. In this light, there is a substantial number of so-called 'international migrants', who had been working abroad as labour migrants in the 1960s, 1970s and 1980s and had returned to Malawi in due course, most of them after the transition to democracy in 1994. These labour migrants also accounted for a small proportion of foreigners in the area, Zambians and Zimbabweans who married Malawians in the migration and had come to Malawi with them.

Although, these days, the bulk of migratory movements are internal (Englund, 2001) and mainly circular in nature (Beegle and Poulin, 2013), a considerable number of the migrants are international migrants, both Malawians who went abroad and foreigners who had migrated to Malawi. Most of them can actually be described as returnee migrants. The return to

Malawi has mostly not been on a voluntary basis but has largely been induced by changing political and economic conditions in their host countries, where they had spent more than 30–40 years of their lives or had been born as second-generation migrants. Although in most countries the economic and social position of Malawian guest workers had been worsening since the 1980s, many had migrated back to Malawi only following the transition to democracy in 1994. Used to a more liberal political climate and better living conditions, many of them had not wanted to return to Banda's authoritarian Malawi. The urban literature of Lilongwe suggests that most international migrants returned to the rural areas (Englund, 2001), yet data from the towns provides a more differentiated picture. Many international migrants came directly to the towns.

As pointed out above, Malawi has become an important destination and transit country for migrants. As a country on the route towards these relatively well-off economies, Malawi also hosts a large number of transit migrants, most of whom are destined to other countries. Its relatively stable political climate and 'permissive' immigration laws have also increasingly made it attractive as a destination country. Countries bordering Malawi have, however, tended to contribute family reunification and economic migrants.

Immigrants to Malawi, as shown in the foregoing analysis, also bring with them business skills that lead to growth in enterprises and employment opportunities. Many migrants have lived in Malawi for years, setting up businesses in the town or marrying Malawians and having children with them. Most of the immigrants from India and Pakistan have well-established businesses in the urban centres of Malawi. Immigration data also shows a growing number of naturalisations from Asian countries, particularly India and Pakistan (Ndegwa, 2015). Burundian and Rwandan immigrants have also emerged as thriving entrepreneurs who tend to operate various retail business, including running hawkers, in townships as well as key trading centres in the city. Younger migrants, often with little or no personal experience of foreign countries, also imagine South Africa and Zimbabwe as attractive avenues to enrichment. While some of them engage in cross-border smuggling from Zambia and Tanzania, others risk considerable sums of money when they travel to Zimbabwe or South Africa to bring goods to Malawi, such as spare parts and electronics. This puts many migrants at risk of human trafficking due to their vulnerability.

Human trafficking is an area of concern in Malawi that was ranked a tier 2 country in the 2000 Trafficking in Person Report, as Malawi does not meet the minimum standard for the elimination of human trafficking activities. Malawi is considered a source, transit and destination country for victims of human trafficking. In 2020, the Government of Malawi identified 140 trafficking victims, of whom 65 were children and 75 adults. Traffickers usually exploit family members from the southern part of the country to the Central and Northern Regions, forcing them to labour in agriculture, goat and cattle herding, begging, fishing and brickmaking, and recently others have been coerced into stealing. Exploitation outside the family involves fraudulent recruitment of people, connected often to physical or sexual abuse. Traffickers typically lure children in rural areas by offering employment opportunities, clothing or lodging, for which they are sometimes charged exorbitant fees resulting in labour and sex trafficking coerced through debts. Malawian victims of trafficking have been identified in South Africa, Kenya, Mozambique, Tanzania, Zambia, Iraq, Kuwait and Saudi Arabia. Criminal networks are active in the Dzaleka refugee camp, where vulnerable men, women and children are being exploited for profits within the camp itself or trafficked into other countries in Southern Africa for forced labour and/or prostitution. The government launched the Standard Operating Procedure (SOP) and the National Referral Mechanisms (NRM), referring all victims to non-governmental organisations where they receive counselling, medical care, food and livelihood training.

Challenges associated with these flows as well as labour migration into and out of the country have created a pressing need for a comprehensive national migration policy.

6.3 Migration Policies and the Regulatory Environment

Policy and legislation have a bearing on migration and are enacted in recognition of the changing global and national development context in which migration and labour issues have to be dealt with. Their enactment is heralding a concerted effort to harmonise different laws touching on migration to ensure that they best capture the interrelation between migration and development.

At the regional level, Malawi is a Member of the Southern Africa Development Community (SADC), handling the movement of people within the region and ultimately removing obstacles to the free movement of goods and services, as well as of capital and labour. Malawi is also a member of the Common Market for East and Southern Africa (COMESA), which adopted the Protocol of the Free Movement of Persons, Labour, Services, and the Rights of Establishment and Residence.

In 1965, Malawi ratified the 1949 ILO Migration for Employment Convention. In 1987, it also adhered to the 1951 Geneva Refugee Convention, the 1967 Refugee Protocol, and the 1975 ILO Migrants Worker Convention. The country ratified the 1969 OAU Convention Governing the Specific Aspects of Refugee Problems in Africa, through which the refugee definition of the 1951 Refugee Convention and the 1969 OAU Convention were incorporated into the country's 1989 Refugee Act. In 1991, the country ratified the 1989 Convention on the Rights of the Child and the 1990 UN Migrant Workers Convention. Malawi is also a signatory to the African Union Convention for the Protection and Assistance of Internally Displaced Persons in Africa. In 2015 Malawi adopted the Trafficking in Persons Act, which criminalises sex and labour trafficking. In 2018, the Government of Malawi agreed to roll out the Comprehensive Refugee Response Framework (GRRF) under the 2016 New York Declaration, which enhances the harmonious co-existence of refugees and asylum seekers with nationals. Despite its enrolment, the government is not implementing the five key pledges made during the GRRF Submit, in order to achieve the goal of facilitating the integration of refugees and asylum seekers within communities through the policy frameworks. Other international migration-relevant protocols ratified by Malawi include (a) 1989 Convention on the Rights of the Child, ratified in 1991; (b) 2000 Human Trafficking Protocol, ratified in 2005; and (c) 2000 Migrant Smuggling Protocol, ratified in 2005.

6.3.1 The Malawi Immigration Act 1964

The Malawi Immigration Act 1964, with its amendments up to 1988, governs the handling of all immigration matters in the country. Under the Act, the legal prescripts governing general immigration and emigration, issuance of residence and other permits and other general migration-related processes are laid out. Part I section 4 (1) of the Act sets out who are regarded as prohibited immigrants on economic grounds, deficiency of education, infirmity of body or mind, mental illness, those suffering from certain prescribed diseases unless issued with a ministerial exemption, convicted felons, grounds of what is regarded as immorality, deportees, wives and children under the age of 18 years and any other dependant relative of a prohibited immigrant, and broadly others who the Minister may deem undesirable based on information received. Part II sections 4 and 5 further set out how officers should conduct medical examination on persons wishing to enter and stay in Malawi for an extended period. The Act is likely to be reviewed once the comprehensive migration and citizenship policy is finalised.

6.3.2 The Malawi Refugees Act of 1989

The Malawi Refugees Act of 1989 established a high-level Refugee Committee with the mandate to hear and grant, refuse or revoke refugee status. The Act further grants the Minister for Internal Affairs broad powers to make further regulations relating to registration, movement and the welfare of refugees, such as relief assistance in cooperation with NGOs. To give effect to the regulation of refugees, the Ministry of Internal Affairs implemented a procedure for determining refugee status. With assistance from the UNHCR, the Ministry of Internal Affairs established a dedicated Refugee Status Determination Unit to deal with eligibility issues.

Malawi acceded to the 1951 Convention relating to the Status of Refugees, as well as its 1967 Protocol (hereafter, 1951 Convention), following accession on 10 December 1989. However, it is one of the countries with reservations on a number of articles in the 1951 Convention. On 4 November 1987, the country ratified the Organisation of African Unity's (OAU) 1969 Convention Governing the Specific Aspects of Refugee Problems in Africa (1969 OAU Convention). The refugee definitions of the 1951 Refugee Convention and the 1969 OAU Convention are incorporated into the country's 1989 Refugees

Act. However, the act was enacted during the Mozambican refugee influx and generally in reaction to refugee management challenges that prevailed at the time it was promulgated. On 7 October 2009, the country ratified the 1954 Convention relating to the Status of Stateless Persons (UNHCR, 2010). The Act was more of a reaction to refugee issues that were relevant to the context then. With the passing of time, a number of gaps have emerged between the law and practice, hence the need for review.

In October 2008, the Government, with funding from the UNHCR, hired a consultant to draft a Refugees Policy and Refugee Amendment Bill. A technical committee was subsequently instituted to liaise with the consultant in the drafting and review process. In 2010, a National High Level Consultative meeting took place in Lilongwe for relevant stakeholders, where the draft Refugee Policy was discussed and adopted.

In September 2011, a meeting of all principal secretaries of the various ministries took place where the Refugee Policy was also adopted. The Bill was also adopted in broad terms but with recommendations of changes. At the moment, both the policy and the draft bill are with the Ministry of Home Affairs, which is mandated to submit them to the Office of the President and Cabinet for further submission to Cabinet.

6.3.3 Employment Act No. 6 of 2000

The labour market in Malawi is regulated under the provisions of the Employment Act No. 6 of 2000. The Act establishes the Office of the Labour Commissioner under the Ministry of Labour, as well as details the duties of the Commissioner and Labour officers to enforce the provisions of the Act. This includes the publication of an annual labour report that must include statistics on the labour market. It is unclear if this is sufficient jurisdiction for the Commissioner to collect data on Malawians leaving the country on employment contracts. Recently, Malawi has entered into a labour export agreement with the United Arab Emirates. Most of those leaving for the United Arab Emirates do so through a private labour recruitment bureau known as the Job Centre. Currently, there is no regulatory or legal requirement for such private labour recruitment agencies to provide data to the Government on Malawians leaving to work outside the country.

The aforementioned instruments need to provide the legal and regulatory framework to ensure that the country fully realises the benefits of migration, and at the same time minimise its pernicious effects. Hence, in recognition of the importance of mainstreaming migration in development, the government should seek to empower Malawians, such as those who have a claim to Malawian heritage, in participating in the development of the country by providing a regulatory environment that explicitly promotes their role through remittances, investments and other socioeconomic activities. This should be counterbalanced with a careful and judicious encouragement of immigrants with desired investment capacity and skills to enjoy safe domicile in the country. At the same time, in cooperation with international and regional partners, there should be concerted efforts to abide by international conventions for the protection of refugees, asylum seekers and other vulnerable migrants while providing clear procedures and practices for dealing with irregular migrants and accompanying law infringement. Migration policies need strong institutions to develop and uphold them at both the local and the international level.

6.4 Migration Institutions and Governance

The frontline national departments dealing with migrants, refugees, asylum seekers and other categories of migrants in Malawi are the Ministry of Home Affairs, the Immigration Department, the Malawi Police Services, the Prison Services, the Ministry of Gender and the Ministry of Foreign Affairs and International Cooperation, and Children, Disability and Social Welfare. Through the Ministry of Home Affairs, the Immigration Department provides services to the general public in areas of border control, issuance of travel documents, residential and work permits, visas and citizenship to eligible persons. The Malawi Police and Prison Services have records of migrants arrested and imprisoned for law infringement. The Ministry of Gender handles cases of minors who are separated, unaccompanied, smuggled and vulnerable women migrants. The Foreign Affairs Ministry deals with the diaspora. Besides the government, international organisations and faith-based organisation also play a critical role in dealing with migrants as outlined below.

(a) Ministry of Home Affairs

Migration management in Malawi is the responsibility of the Ministry of Home Affairs and Internal Security, which also includes the Immigration Department and Department of Refugees. The Ministry of Home Affairs, through its Refugee Unit, also handles all refugees and asylum-seekers status determination, working very closely with the UNHCR, who also have data on refugees in Dzaleka and those in the Karonga transit centre. The United Nations High Commission for Refugees (UNHCR) in Malawi keeps well-documented and detailed data on refugees and asylum-seekers. The UNHCR, in partnership with the Ministry of Home Affairs' Refugee Unit, handles all refugees and asylum seekers. The Ministry also carries the legislative mandate on migration and internal security affairs. Recently, following the just concluded tripartite elections, the Ministry was also charged with handling the civil registration docket, which is responsible for issuing vital registration documents, such as birth and death certificates. It will also be responsible for issuing of soon-to-be introduced national identity documents for Malawi nationals.

(b) Immigration Department

The Immigration Department exists under the Ministry of Home Affairs. The overall responsibility for institutional management and development lies with the Chief Immigration Officer, who is the head of the department. The department is governed by the Immigration Act 1964 (including amendments up to and including 1988), the Citizenship Act of 1966 (including amendments up to and including 1972) and the Laws of Malawi and the Republic of Malawi Constitution under section 47.

The department manages people who are entering and leaving the country in order to uphold the security of the State. It also issues travel documents to eligible persons in accordance with the International Civil Aviation Standards. The department also processes and issues Malawi Citizenship, Residence and Work Permits and visas in accordance with the existing policies. The department also has mandate to monitor, track, apprehend and repatriate illegal immigrants according to the Laws of Malawi and existing policies.

(c) **Department of Refugees**

Also under the Ministry of Home Affairs, the Department of Refugees is responsible for managing refugee affairs in the country and coordinating the refugee programmes in collaboration with the UNHCR and other partners. It derives its mandate from the Malawi Refugees Act and other international instruments, such as the 1951 Convention and the 1969 OAU Convention. Its main tasks are the formulation and review of the Refugee Policy and the verification exercise of the population of refugees and asylum seekers in the country.

(d) **Malawi Police Service**

The Malawi Police also falls under the Ministry of Home Affairs and Internal Security. It handles all detentions and arrests of migrants who infringe the law. This applies to those who are suspected of infringing the Immigration Act, as well as other aspects of the law in the country. To ensure that the real law breakers – such as smugglers, traffickers and those who exploit vulnerable migrants – are arrested, IOM is working with the Government of Malawi to train Immigration officials on the need to decriminalise migrants.

Currently, because the Immigration Department does not have detention facilities for suspects, all immigrants suspected of infringing the immigration law are detained in police cells together with other law breakers. This is in violation of the UNHCR guidelines (UNHCR, 2012) on the use of appropriate alternatives to detention for refugees and asylum-seekers. The recommendations emphasise the right to seek asylum and the right to liberty. To ensure that persons are not deprived of their liberty arbitrarily, detention must be in accordance the national and international law. Furthermore, the guidelines indicate that detention must be based on the assessment of an individual's circumstances. Because detention is an exceptional measure, the guidelines indicate that detention can only be justified for a legitimate purpose. Some of the purposes may include protection of public order, to protect public health or to protect national security.

(e) Ministry of Gender, Children, Disability and Social Welfare

The Ministry of Gender, Children, Disability and Social Welfare handles all cases of minors who are separated, unaccompanied, smuggled, those who have been legally adopted and also deals with women migrants in vulnerable circumstances. It also collects and reports on such cases and determines forms of social assistance to be provided to migrants, refugees and asylum seekers. It handles cases of repatriation to and from Malawi, as well as provides assistance to children who travel from Malawi after being adopted. It is also responsible for child protection involving cases of non-Malawian children in the country and Malawian children outside the country. The Ministry offers various forms of assistance, ranging from providing supporting documentation to those seeking visas after adoption, to counselling for refugees and asylum-seekers in Malawi.

The Ministry has district offices that receive and deal with cases at the district level. Only few cases are handled at the national level and data protection is not collated at the national level from the districts. This makes it difficult to discern patterns and trends from existing records, which are also rarely converted into electronic data format.

(f) Ministry of Foreign Affairs

The Ministry of Foreign Affairs deals with migration to the extent that there are Malawians outside the country and at the bilateral level with respective countries regarding their nationals who may be visiting Malawi. Recently, the Ministry established a unit to handle diaspora affairs. The unit works through Malawi missions abroad to collect information on Malawians outside the country. It is currently in discussion with the Immigration Department to enhance their data collection on Malawian diaspora. Through Malawi missions abroad, the Ministry also provides consular services and trade and investment activities for the country.

(g) Ministry of Labour

The Ministry of Labour holds the main mandate for negotiating and concluding bilateral labour agreements, such as the most recent one with the United Arab Emirates. It also formulates and enforces labour laws in the country, such as those related to the employment of foreign workers. The Ministry also collects data on trends in the labour force for the country.

As pointed out earlier, labour migration and credible estimates of Malawian diaspora and the remittances they send back to the country are scarce.

(h) National Statistical Office

The NSO is cited in the MGDS II as a key agency that monitors, evaluates and tracks progress towards the realisation of the country's social and economic development goals. This is in recognition of its mandate and central role as the official agency for the collection and dissemination of statistics in the country.

The NSO publishes an analytical report on migration data. In addition to the migration report, other thematic reports included "evaluation of the census data quality; population structure by sex and age; spatial distribution of the population (including urbanization); economic characteristics of the population; fertility and nuptiality; mortality; household and living conditions; population projections; women in Malawi; children and youth; population with disability; elderly; and literacy and education ... demographic atlas; district monographs and poverty maps" (NSO, 2010).

(i) Reserve Bank of Malawi

The RBM's role in migration affairs relates to its statutory mandate to capture investment and trade-related inflows into and outside the country. It is also supposed to capture and track remittances to and from Malawi. However, it is hamstrung by the lack of legislation that gives authority to compel money-transfer agents and financial institutions to provide data on funds sent outside and into the country. The bank is working towards putting in place regulatory requirements to this effect.

The Government of Malawi also demonstrates awareness of the critical role of institutions, such as the NSO, international partners including donors, UN agencies and other non-state development agencies in partnering to achieve desired development objectives. In terms of operational handling of migrants, refugees, asylum seekers and other categories of migrants in Malawi, the main front-line agencies that are involved include the government ministries of Home Affairs, Immigration Department, Malawi Police Services, Prisons Services, Ministry of Gender, UNHCR, and more recently, the IOM.

(j) International Organisations and other organisations

The key international organisations dealing with migration-related issues include the International Organisation for Migration (IOM) and the United Nations High Commissioner for Refugees (UNHCR).

Since 2014, the IOM has been very active in Malawi. It has contributed enormously to the Government of Malawi's effort to manage migration through a large variety of projects and programs, which include, among other actions, helping the government to develop the first-ever migration profile, assisting in the voluntary return and reintegration of Malawians, refugee resettlement, migration and health, combating trafficking in persons and human smuggling. The UNHCR supports the government in the refugee response plan. Some of the UNHCR's main activities assisting refugees in Malawi are in the areas of education, health, community empowerment and self-reliance. Other migration-related United Nations agencies in Malawi include Plan International that advances children's rights and equality for girls including refugees, the World Food Programme (WFP) that works to achieve and maintain food security among refugees, transitioning from monthly food distribution to cash-based transfers, and the Moravian

Church that is present in the Dzaleka refugee camp, providing, for example, food relief parcels.

'There is Hope' is a locally based NGO, which through education scholarships, advocacy, social enterprise, leadership development and vocational training, provides assistance to refugees and members of the host communities to overcome poverty and become self-reliant. Through advocacy and lobbying, civic education, training, capacity building, networking and research, the Centre for the Development of People (CEDEP) creates a legally and socially accepting environment where minority groups like refugees, asylum seekers and migrants have an improved livelihood in Malawi.

(k) Faith based Organisations

There are currently eight dioceses in Malawi: two archdioceses in Blantyre and Lilongwe, and six dioceses in Chikwawa, Deza, Mangochi, Karonga, Mzuzu and Zomba. The Catholic Church is committed to providing social services to all who live in Malawi, especially in educational institutions and hospitals. From a non-partisan perspective, the Malawian bishops played a pivotal role in ushering the democratic rules in the country, thanks to their pastoral letter 'Living our Faith' to promote the development of a representative democracy that Malawi enjoys today.

The bishops have entrusted their pastoral services to the Catholic Commission for Justice and Peace (CCJP), which has been very active in the Dzaleka refugee camp. However, since 2019 the Commission has not done any specific activity on behalf of refugees and asylum seekers.

The Catholic Church runs several projects that cater to the needs of refugees, asylum seekers and IDPs in collaboration with various organisations, like the Jesuit Refugee Service (JRS), the Catholic Development Commission (CADECOM) and the Catholic Relief Service. In Malawi, the JRS educates more than 5,000 children in the Dzaleka refugee camp in the pre-primary, primary and secondary schools. The JRS also offers pastoral and psychosocial programmes to refugees requiring support for mental traumas and other livelihood programmes for adults. In a move to empower refugee girls, increase their access and improve the quality of their education, security and overall well-being, JRS Malawi launched the Naweza *'I CAN'* Project. In its educational drive, the JRS and other partners also offered a digital training program in the Dzaleka refugee camp, allowing students to participate in online classes and ensuring the safety of partners and participants. In the wake of the coronavirus outbreak in Malawi and because of the detection of cases in the camp, the JRS has promoted awareness campaigns and, together with partners, has mounted water points and placed hand soaps in strategic areas of the camp.

Apart from the JRS, the Society of Jesus (Jesuits) is also addressing the issue of environment and climate change. The Jesuit Centre for Ecology and Development (JCED) in Malawi runs its programmes in Lilongwe and Kasungu. In these two places, Jesuits work with local communities and promote the protection of the natural environment, as well as the creation of sustainable livelihoods. Even though the JCED does not work directly with the environment of refugees or displaced people, its projects aim to build resilient and sustainable communities. By doing so, the JCED prevents people that depend on agriculture to be particularly vulnerable to the effects of climate breakdown. Thanks to its activities in the country, the JCED won the 2020 Misean Cara Climate Action Award.

Natural disasters, for example, the Tropical Cyclone Idai that ravaged parts of Malawi, caused the internal displacement of people, with subsequent food, water and shelter shortages. The CRS, with other partners in Malawi, worked quickly and provided necessary support to those being affected. Currently, the CRS is offering food security programs in partnership with the US Agency for International Development (USAID), emergency response, water, hygiene and sanitation (WASH) assessments, and capacity-building activities funded by public and private donors throughout the eight Malawian dioceses.

Caritas was established in Malawi in 1984. It was renamed CADECOM (Catholic Development Commission) in 1999, changing its image from a relief organisation to a development agency; however, it is still a member of Caritas Internationalis. Its mission is to create awareness and empower disadvantaged men, women and youth through a participatory development-driven approach. Its programs involve disaster risk reduction, integrated food security, water and sanitation projects.

Migration policy ought to be informed and strengthened by evidence drawn from migration statistics. Therefore, there is an urgent need to improve migration data to comprehensively inform policy through several efforts, such as addressing the gaps in migration data, formulating a national migration data management strategy and facilitating regular reports on migration, including regularly updating the migration profile. There are also shortfalls in the coordination and cooperation among departments that deal with migration issues, particularly with regards to the collection, analysis and dissemination of migration data. As a result, migration data is scattered between ministries, departments and, in some cases, not captured at all. This makes it difficult for policymakers to tap into relevant sources of information and data to inform policy.

6.5 Summary

Malawi is one of the impoverished countries in the region. However, despite its economic status, it does have its own share of immigrants – some who come legally looking for employment and investment opportunities and others illegally cross its borders. Its institutional framework for migration, though functional, is in a dire need of revision. As a signatory of international and regional organisations, its migration policies and legislations are partly informed by these protocols. The country receives a lot of refugees and illegal migrants from Central and North Africa who normally use it as a transit point to other countries in the region (such as South Africa and Zimbabwe) and abroad. Its stable political environment, coupled with its permissive immigration control policies, makes it a country of choice for those seeking better opportunities in the country or beyond.

Notes

1 https://www.un.org/en/development/desa/population/migration/publications/wallchart/docs/MigrationWallChart2017.pdf
2 https://data2.unhcr.org/en/country/mwi

Bibliography

Beegle, K. & Poulin, M. 2013. Migration and the Transition to Adulthood in Contemporary Malawi. *The Annals of the American Academy of Political and Social Science*, 648, 38–51.

Callamard, A. 1994. Refugees and Local Hosts: A Study of the Trading Interactions between Mozambican Refugees and Malawian Villagers in the District of Mwanza. *Journal of Refugee Studies*, 7, 39–62.

Englund, H. 2001. The Politics of Multiple Identities: The Making of a Home Villagers' Association in Lilongwe, Malawi. In *Associational Life in African Cities: Popular Responses to the Urban Crisis*, 90–106. Uppsala: Nordic Africa Institute.

Kainja, G. 2012. A Call for a Multi-Sector Approach Against People Smuggling/Illegal Migration: A Case Study of Malawi. Research Paper, Research and Planning Unit, Malawi Police Service, Lilongwe, Malawi.

Kloukinas, P., Novelli, V., Kafodya, I., Ngoma, I., Macdonald, J. & Goda, K. 2020. A Building Classification Scheme of Housing Stock in Malawi for Earthquake Risk Assessment. *Journal of Housing and the Built Environment*, 35, 507–537.

Manda, M. 2013. *Malawi Situation of Urbanisation Report*. Lilongwe, Malawi: Government of Malawi.

National Statistical Office. 2010. Malawi Demographic and Health Survey (2010 MDHS). National Statistical Office, Zomba, Malawi and ICF Macro, Calverton, Maryland, USA.

Ndegwa, D. 2015. *Migration in Malawi: A Country Profile 2014*. International Organization for Migration.

Nkhoma, B. G. 2014. Transnational Threats: The Problem of Illegal Immigration in Northern Malawi.

Potts, D. 1986. *Urbanization in Malawi with Special Reference to the New Capital City of Lilongwe*. University of London.

Rust, K. & Gavera, C. 2013. Housing Finance in Africa: A Review of Some of Africa's Housing Finance Markets. Johannesburg, SA: Center for Affordable Housing Finance in Africa.

Un-Habitat. 2010. Malawi Urban Housing Sector Profile, 131. https://issuu.com/unhabitat/docs/malawi_urban_housing_sector_profile

United Nations. 2017. International Migration 2017. Department of Economic and Social Affairs: Population Division. Accessible online: https://www.un.org/en/development/desa/population/migration/publications/wallchart/docs/MigrationWallChart2017.pdf

7 Maputo
The Lush Capital of Mozambique

7.1 Introduction

Maputo, formerly (until 1976) Lourenço Marques, is a port city and the capital of Mozambique. It lies along the north bank of Espírito Santo Estuary of Delagoa Bay, an inlet of the Indian Ocean at the confluence point of four rivers. Settlement in the area of present-day Maputo initially began in the 1500s as a fishing village. In 1781, the Portuguese established a fort in the area which was named Lourenço Marques after the navigator who had first explored the area in 1544. The town developed around a Portuguese fortress completed in 1787. By 1850, a settlement that fit the description of a town had grown around the fort. In 1877, the town was elevated to the status of a city, and in 1898 it became the capital of the Portuguese Mozambique colony. The Maputo province surrounds the city, which is a separate province by itself. The city and its port continued to grow and flourish into the 20th century. Finally, it became the capital of the independent nation of Mozambique in 1975 upon Mozambican independence. After Mozambique's independence from Portugal, the city was first renamed Cam Phumo, its pre-colonial name, and then Maputo in 1976. The name comes from the Maputo River, which marks a part of the southern border of the country.

Today, Maputo is an African cosmopolitan metropolis. Together with its suburb Matola, it forms Metropolitan Maputo, a city with a population of 2.5 million people. The spoken languages are Portuguese (official) and various Bantu languages, but also Arabic, Indian and Chinese. Maputo City is the nation's most populous city and Mozambique's smallest province by area but the most densely populated one. Maputo has a total area of 346.77 km^2. It has a population of around 1,273,076 and a population density of 3.648 inhabitants/km^2.

It is a port city with a commerce-based economy. The port is one of the most important ones in East Africa, which before independence handled transit trade from the mines and industries of South Africa, Swaziland and Rhodesia, with which it has rail and road connections. After the frontier with Rhodesia was closed, and as Mozambique–South African relations became increasingly strained, the port suffered. Local industries include brewing, shipbuilding and repair, fish canning, iron working and the manufacture of

DOI: 10.4324/9781003184508-7

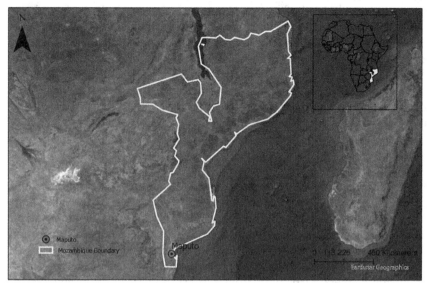

L. Chipungu

Figure 7.1 Map of Mozambique Showing the Position of Maputo

cement, textiles, and other goods. It has an economy that is centred around its port. As the seat of the national government of Mozambique, Maputo houses all the important government offices, ministries, official residences, embassies, etc. The Parliament of the country meets here, and the government operates from the city. Maputo is also a melting pot of many cultures and, as such, hosts numerous cultural attractions and institutions.

Forty per cent of the urbanisable spaces is occupied by areas that, despite the lack of infrastructure, have formally demarcated plots. The lack of building control is leading people who are occupying these spaces to densify the number of families per plot. The land considered as Developed Urban Areas and Developing Areas correspond, respectively, to the consolidated areas of the Municipality and to areas still susceptible to densification. Industrial activities, storage facilities and workshops occupy approximately 456 hectares. Land consumed by agricultural practices, still one of the main livelihood activities for a considerable part of the population, occupies 25% of the municipal territory.

The Municipality of Maputo, whose area is 347 km^2, has an urbanised area of about 117.6 km^2 (excluding the areas as forest, indicated in the current land use plan). Thirty-three per cent of Maputo's municipal land is dominated by residential uses. The population density in developing urban areas/zones has on average of 70 inhabitants per hectare. About 14% of the population are concentrated essentially in the Municipal District N°1. In developed urban areas (8,400 ha), the population density is characterised by single-family residential typology with a lack of infrastructure. There are extensive areas in need of proper planning, i.e. areas that lack of legal certainty of use, demarcation and registration in the municipal register, a situation that makes it difficult for users to access services such as water and energy distribution networks, rainwater runoff. The population of Mozambique's capital city, Maputo, continues to grow at a rapid rate as the result of high birth rates and immigration, posing enormous challenges to the local government in its efforts to deliver basic services, provide food and improve the city's infrastructure.

7.2 Migrant Profile in Maputo

With 921,513 citizens living abroad (12%), Mozambique is the primary country of origin of immigrants in the Southern African area. Historically, many people migrated from Mozambique to South Africa to work in mines and on farms owned by corporations, but more recently, internal labour migration has increased as the nation has opened up to mining and energy businesses. Additionally, as Mozambique's economy continues to expand, more people are migrating there, particularly from the north and centre of the country, for work in the country's mines or as a stepping stone to South Africa. Zimbabwe (around 100,000), Malawi (60,000), Angola (40,000), Kenya (25,000) and South Africa (20,000) are the top five countries of origin.

By the middle of 2020, there were 338,900 international migrants and 640,200 emigrants. A five-year period ending in 2019 saw little change in the migration trend, with a net movement of −25 000 and slightly more emigrants than immigrants. About 52.1% of immigrants in 2020 were female, 26.8% were under the age of 19 and 3.2% were 65 years or older.

Maputo is home to migrants from neighbouring countries. In the 2017 census, 2% of the city's population was recorded as foreign nationals (Instituto Nacional de Estatísticas, 2019).[1] South Africans were the most common nationality, with a large number of young immigrants coming to Mozambique in search of work in mines, plantations or entrepreneurial activities (ODI and CESC, 2011).[2] A similar number of Portuguese immigrants was also recorded. Mozambique's location, linking landlocked countries such as Zimbabwe, Zambia and Malawi to major seaports and to South Africa, makes it a significant transit country. Maputo has two high traffic transport corridors, connecting the city with Johannesburg in South Africa and Mbabane in Eswatini.

Mozambique currently hosts persons of concern, including refugees and asylum seekers. At the end of January 2021, Mozambique was home to about 27,193 refugees and asylum seekers, 33% of whom reside in the Maratane settlement in the Nampula Province and the remaining 67% of forced migrants who do not choose to stay in the refugee camp self-settle in urban areas, namely Maputo (the country's capital) and Nampula City. The majority of these refugees and asylum seekers were from the Democratic Republic of the Congo (32.1%), followed by Burundi (28.2%), Rwanda (12.3%), Somalia (12.1%) and Burundi (12.2%), with 15.2% coming from other nations. Migrants who reside in urban areas are generally self-reliant, thanks to Mozambique's favourable protection environment. The vast majority of refugees and asylum seekers in Mozambique have transited through refugee camps in other countries, such as Tanzania, Zambia and/or Malawi, in order to settle in Mozambique. Around 45% of the refugee and asylum-seeker population in Mozambique are females and over half of the population is below the age of 18 years.

Mozambique is attractive to refugees largely due to the relatively beneficent position taken by the government towards refugees as compared to other countries in the Southern Africa region, since refugees are allowed to work, have free access to all markets and enjoy freedom of movement (Malauene and Landau, 2004). Once asylum seekers receive refugee status, they enjoy the same socio-economic rights of a foreigner residing in the country, among which are the rights to education and health care, and have to respect and observe the Mozambican laws, including any instructions relating to the maintenance of public order, and should abstain from any subversive activities against the state.

Access to decent and affordable housing is a fundamental human need and a human right. Access to medium- and long-term housing solutions in the years immediately following migrants' arrival in the host country is a critical element in their long-term integration prospects, including access to employment and education, among other issues. It is also a precondition for newcomers' integration into a new society. In Maputo, most migrants access housing or accommodation in buildings or garages transformed into rooms, mostly in the Malhangalene, Central and Alto Maé neighbourhoods, where they live mostly with friends from their country of origin and members of their pre-departure households. Forced migrants pay higher rent costs compared to the amounts paid by local Mozambicans, just because they are foreigners and perceived to have more money. To overcome accommodation constraints and considering that they do not wish to spend more than necessary on accommodation and furniture since they can leave at any moment, it is common to find from three to seven migrants sharing a single room. Others who lack money for rent depend on friends for lodging. In case of delay in rent payment, the landlords' attitude towards migrants is very rude, with some forced migrants having their goods thrown outside the apartments and being insulted by landlords in front of their neighbours.

Lloyd-Jones and Rakodi (2014) and Majale (2001) reveal that in urban settings it is difficult for the most vulnerable (a category in which forced migrants are included) to pursue sustainable livelihoods, because they have constraints accessing labour; they lack a robust social network that could be effective in providing support and assistance; they are unable to access adequate housing, which denies them the opportunity to earn livelihoods through home-based entrepreneurial activities; they face problems accessing credit in the formal financial sector; and, if they find employment at all, they depend on their labour for income that is needed for daily needs such as to pay for rent, transportation, food and health care, which increases their living costs. This situation is worsened for self-settled forced migrants considering the additional challenges they face such as their legal status, difficulties getting a work permit, etc., which leads them to the informal sector, starting businesses or working for small businesses. Although they have those livelihood strategies, refugees often remain more marginalised since they lack the rights that accompany citizenship, rights that in some countries even the national citizen do not enjoy.[3]

With regard to accessing basic services such as health care, the interviews conducted with diplomats in this study reveal that challenges faced by migrants in accessing health care services are moderate. For forced migrants, the UNHCR promotes refugees' inclusion into the national health system, including referral mechanisms, sexual and reproductive health rights (SRHR) services, prevention and treatment of HIV and mental health and psychosocial support. To prevent COVID-19 in the Maratane refugee camp, the UNHCR has conducted information campaigns with refugees to promote frequent handwashing, social distancing and the use of facemasks. Additionally, the UNHCR has conducted regular distributions of soap, and assisted refugees in producing face masks as part of a strategy to prevent COVID-19 while simultaneously stimulating livelihoods opportunities (UNHCR, 2020).

While self-settled in urban areas, it is expected of migrants to integrate into the host community, to connect with the host population, to support themselves and their families and to network socially. Contrary to these expectations, in Maputo City some migrants, especially forced migrants, tend to not integrate with the host community. They live in a state of 'permanent transit', in which they stay in urban settings and live for a couple of years (around 3–5 years) but continue to see their situation as temporary, holding on to hopes of resettlement, and waiting to leave at any moment for a third country of asylum. They are physically living in Maputo City, but their focus and futures are presumed to be elsewhere, because they do not want to stay there. Because of this 'permanent transit' condition, forced migrants, especially Congolese, tend to not get attached to Mozambicans through marriage and/ or children; they often do not get jobs despite being qualified to do so and tend to engage in socio-cultural activities with other Congolese rather than with Mozambicans. Malauene and Landau (2004) states that, while in exile, refugees tend to recreate their communities: they tend to respond to diversity by sticking together, as manifested by their preference to settle in the same geographical area and as far as possible to interact almost exclusively with their own ethnic groups. Forced migrants also use their rich cultural heritage – music and dance – to try to rebuild their cultural identity in Mozambique.

In relation to the sense of belonging, few migrants feel part of the Mozambican society, while at the same time many are proud of identifying as citizens of their country of origin. They do not feel a part of the Mozambican society because they do not enjoy the same rights as the locals. Furthermore, most migrants do not view themselves living in Mozambique in the future or teaching or bringing up their children according to the Mozambican culture. Some even want their children to be taught in English in a Portuguese-speaking country.

Despite reservations to the 1951 Convention, refugees and asylum seekers have access to education, as well as the right to work in Mozambique. In the Maratane refugee camp, there is one primary school and one secondary school run by the Ministry of Education with support from the UNHCR. The schools serve both refugee and host community children to promote social cohesion and peaceful coexistence. In Maratane, there are 2,757 children enrolled in primary school (1,489 refugees and 1,268 from the host community) and 713 students enrolled in secondary school (501 refugees and 212 members of the host community). At present, the UNHCR is working with the Vodafone Foundation and the Ministry of Education to implement the Instant Network Schools (INS) programme in secondary schools in Maratane and Nampula. The INS is an integrated platform that transforms an existing classroom into an innovation hub. Moreover, refugees have access to tertiary education under the same conditions as nationals, although the government does not provide higher education scholarships to foreigners. Nonetheless, since 2019, 34 students have accessed higher education through the DAFI scholarship programme (Albert Einstein German Academic Refugee Initiative) with the support of the UNHCR (UNHCR, 2020).

Notwithstanding, forced migrants in 'permanent transit' condition are predominantly unemployed. Some migrants are studying and survive with remittances from friends, relatives and family in the country of origin or in the wider diaspora. They receive money monthly through Western Union in the Mozambique International Bank (BIM) that charges 20% tax fee. To access the money, migrants must show the identity declaration and the transfer code. Remittances for migrants are eased by their transnational networks. They use telecommunication centres to establish contact with relatives back home or in the wider diaspora, and to organise visits, money transfers, get updated news and sometimes to inform about financial needs and constraints.

In Maputo, they live for 3–5 years, waiting to leave at any moment for a third country of resettlement; they are physically living in Maputo, but they are theoretically travelling abroad. They do not want to stay in Mozambique, and they hold on to hopes of resettlement, consequently resisting integration or any form of settlement that might jeopardise or obstruct resettlement opportunities.

Migrants of African origin want to go to Western countries for economic and welfare betterment, job and education opportunities, and to join their relatives, friends and family. Some even try South Africa but eventually return to Mozambique for fear of crime and xenophobia. There is also a perception that in Mozambique it would be easier to get passports and visa to European countries, the United States of America and Australia.

Local integration means that the host government offers permanent asylum to refugees and full integration into the host society. The host government grants membership and residency status to refugees. Local integration takes place through a process of legal, economic, social and cultural incorporation of refugees, culminating in the granting of citizenship (Kibreab, 1999).

As referred by Jacobsen (2001), refugees with this status enjoy a range of human and civil rights, which includes the right to marry, to practice one's own religion, to own property, to work and seek employment, to have access to education and housing. However, in reality, despite the insecurity of their legal status and the temporariness of their stay, over time refugees integrate unofficially into host communities (Jacobsen, 2001).

Migrants and refugees in Maputo face many obstacles, while at the same time having a great deal of opportunities at their disposal. As Miamidian and Jacobsen (2004) observe, the main obstacles faced by migrants include legal documentation, resistance from local communities, resistance from law enforcement agencies, government policy towards migrants, especially refugees, cultural issues and language barriers. Most forced migrants have lived in Maputo for some years without refugee status, holding only a declaration issued by the Mozambican authorities as an identification document. Documentation is a huge constraint for forced migrants. Due to the lack of proper documentation, forced migrants report facing problems accessing employment, government social grants, bank accounts, financial credit, education and restricted movement and psychological fear of arrest as they frequently must bribe police officers to avoid going to jail. On the other hand, urban refugees' skills, business knowledge and knowledge of other countries' markets gives them an advantage to undertake import–export activities and to grow micro enterprises into small businesses (Miamidian and Jacobsen, 2004).

Regarding the ability to sustain livelihoods, since forced migrants in urban settings only get legal assistance from the UNHCR and depend on other non-refugee related means of survival, it is expected that they find employment. In general, forced migrants' employment status in Maputo is diversified. The activities by those employed full time include teaching French and English languages, voluntary work in NGOs and working as locksmiths, police officers, microfinance officers, electricians, doctors, lawyers and mechanics. Forced migrants can access jobs due to an exception created by the Ministry of Labour to allow them to work, despite the legal indication that they can only work in Mozambique when holding the refugee status (Miamidian and Jacobsen, 2004). Migrants are employed by Mozambicans and other immigrants, probably from the same tribe or ethnic group or country.

Some migrants in Maputo are self-employed, running small businesses (barber shop, selling groceries) and doing occasional work. According to Jacobsen (2004), urban migrants often bring with them new or different skills, knowledge of markets in their home countries, and are more experienced in business than their local counterparts, which constitute a competitive advantage for the Mozambican business environment. The lack of access to social welfare, family support and formal employment opportunities means that migrants are more willing to take financial risks to make their businesses prosper.

There are many more opportunities in Mozambique, but it depends on your area of expertise. If you're an architect or engineer, or have technical skills, there are plenty of jobs. There has been a 30–40% increase in the number of Portuguese migrants choosing to move to Mozambique over the past two years. As Portuguese nationals don't have to register at the embassy, concrete numbers are hard to come by. Many migrants arrive on fixed-term contracts with Portuguese companies who have invested in Mozambique.

Many Portuguese in Maputo run small- or medium-scale enterprises. Often these are restaurants, cafés or smaller specialised companies in the construction sector. In the last category, there are consultancy companies offering expertise in engineering and quality control; specialised building companies carrying out, for instance, electric installations; and companies providing imported building material. Both Mozambicans and Portuguese tend to see *hotelaria* (hotels and restaurants) as well as construction as sectors of Portuguese expertise. The social relationship between Portuguese employers and Mozambican employees has its roots in colonial times, and generally is still characterised by 'abyssal' (de Sousa Santos, 2018) power inequalities, precluding any form of social equivalence between the two.

7.3 Migration Policies and Regulations

Mozambique is a party to the 1951 United Nations Convention relating to the Status of Refugees (ratified on 16 December 1983), the additional 1967 Protocol (ratified on 1 May 1989), as well as the 1969 Organization of African Unity Convention governing Specific Problems of Refugees (ratified on 22 February 1989). Despite these reservations, the UNHCR is satisfied that Mozambique generally maintains a generous asylum policy through the adoption of practical arrangements, which grant asylum seekers and refugees rights similar to those of its nationals. Therefore, these limitations have had so far little impact on the actual treatment of refugees and asylum seekers. The only restriction is the limitation of the right of refugees and asylum seekers to freedom of movement and choice of residence, based upon a Ministerial Instruction, issued in 2001 and implemented in 2003, banning refugees from residing in the Maputo capital. Aside from Refugee Legislation there is the Nationality Act 2 of 1982 and the immigration legislation in the form of the National Assembly Law 5 of 1993, which governs and 'establishes juridical regime for foreign citizens namely, norms of entry, residence and departure from the country, rights, duties and privileges while in the country'(Migrants and Refugees, 2021).

Mozambique is rarely the first place of asylum for refugees, since most of them have passed through at least one other country before arriving in Mozambique. Since Africa does not have a European-style first-country-of-asylum system, Mozambique accepts asylum seekers, regardless of whether they have resided in another country before arriving in Mozambique. However, legally, asylum seekers who have already received refugee status in another country might not be eligible for refugee status in Mozambique (Malauene and Landau, 2004).

- **Mozambique: Act No. 21/91 of 31 December 1991 (Refugee Act)**[4]
  Through the appropriate instrument, the Republic of Mozambique acceded on 22 October 1983 to the Convention relating to the Status of Refugees of 28 July 1951, at the same time making its reservations in accordance with Article 42 of the said Convention.

  The Mozambican State, by resolutions Nos. 11/88 and 12/88 of 25 August, ratified the Convention of the Organization of African Unity governing the specific aspects of refugee problems in Africa, dated 10 September 1969, and the additional Protocol to the Geneva Convention relating to the Status of Refugees, dated 31 January 1967.

The Refugee Act as approved by the Republic Assembly on 31 December 1991 (law 21/91), refugees' rights and obligations, and the competencies in the refugee determination process are defined. Isen and Halperin (2003) describe how the refugee determination process works:

> Mozambique does not have a prima facie status determination process; all cases are determined on an individual basis. In addition, Mozambique does not screen asylum seekers at its borders. When refugees arrive in the country, they are invited for an interview, after which they pass in front of an 'Eligibility Commission', which decides on a case by case basis whether they are eligible. If the asylum seeker is eligible, he or she is transferred to the 'Minister of Home Affairs' who will decide whether or not to grant the applicant refugee status. If the case is rejected by the Minister, the asylum seeker is given a second chance, after which he or she becomes illegal if they remain in the country.
>
> (Malauene and Landau, 2004)

For the purpose of correct implementation of the above-mentioned Conventions and Protocol, it is necessary to establish the appropriate procedural mechanisms to guide all the formalities with which applications for refugee status for the persons concerned should comply. The legal establishment of these mechanisms reflects an activity forming a necessary complement to the contents of the above-mentioned Conventions, both to ensure legality in the application of the said instruments and to make it possible to conform to the same legality in relation to petitions for asylum, from the submission of the relevant petition up to the final decision thereon, the ultimate objective being to apply the constitutional principle of respect for and defence of human rights.

Wherefore, pursuant to Article 135, paragraph 1, of the Constitution, the Assembly of the Republic hereby decides as follows:

Article 1 (The concept of the refugee)

1 Any person:
   a) Who has a well-founded fear of being persecuted for reasons of race, religion, nationality, membership of a particular social group or

political opinion, is outside the country of his nationality and is unable or, owing to such fear, is unwilling to return to that country or to seek its protection;
b) who, not having a nationality and being outside the country of his former habitual residence, is unable or, owing to such fear, is unwilling to return to it;
c) Who, owing to external aggression, occupation, foreign domination or events seriously disturbing public order in a part or the whole of the country of origin, is compelled to leave his place of habitual residence in order to seek refuge in another place outside his country of origin or nationality shall be considered a refugee.

2 A person who has more than one nationality shall be eligible for refugee status only when the above-mentioned grounds apply in relation to all the States of which he is a national.

Article 5 (Juridical status of the refugee)

1 The refugee shall in principle enjoy the rights and have the duties proper to aliens resident in the Republic of Mozambique; he shall essentially be bound to conform to and obey the legislation in force in the country, including any instructions relating to the maintenance of public order, and to refrain from any subversive activities against any foreign State.
2 The refugee shall enjoy any rights not applicable to aliens in general which arise out of the United Nations Convention of 28 July 1951, the additional Protocol thereto of 31 January 1967 and the OAU Convention of 10 September 1969, subject to the reservations made by the Republic of Mozambique.
3 An identity document attesting his refugee status and, when he has to leave the country, a travel document shall be issued to the refugee.
4 The documents referred to in the previous paragraph may be withheld on grounds of national security or public order determined by the Republic of Mozambique.

Article 12 (Naturalisation)

1 The Republic of Mozambique may authorise the acquisition of Mozambican nationality by naturalisation for any person who has refugee status and who seeks to acquire such nationality by that means.
2 Once the requirements of the legislation concerning nationality have been met, naturalisation shall be granted on the same terms as to other aliens.

Despite the reservations registered, it is pleasing to note that Mozambique in general maintains a generous asylum policy through the adoption of practical arrangements which grant asylum seekers and refugees rights similar to those of its nationals. Reservations to the 1951 Convention have hence had limited impact on the actual treatment of refugees and asylum seekers, who enjoy most of the relevant rights in practice. The only restriction observed is the limitation of the right of refugees and asylum seekers to freedom of movement and choice of residence, based on a Ministerial Instruction, issued in 2001 and implemented in 2003, banning refugees from residing in the capital city of Maputo. This has been applied to those who had not settled in Maputo before 2003. Nonetheless, lifting of the reservations remains an important priority in order to establish an enabling and durable protection environment for the local integration of refugees.

- **Mozambique: Immigration Law No. 5/93 of 1993 (aliens)**
  In order to define the legal framework relating to the management and implementation of migration and to reflect in the legal system the advances arising from international conventions, the government approved Act No. 5/93 of 28 December 1993, which establishes the legal system governing foreign citizens with respect to their entry into, stay in and departure from the country, and to rights, duties, privileges and guarantees while in the country. The applicability of this legal regime for the citizen does not affect the established special laws, bilateral or multilateral accords or International Conventions to which Mozambique is a signatory.

  Article 4 (Rights, Duties and privileges of Foreign citizens) of Act No. 5/93 stipulates the following general principle:[5]

  1. A foreign citizen resident or temporary resident in the country has the same rights and privileges and is subject to the same duties as a Mozambican citizen.
  2. Duties of a foreign citizen in the country are as follows;
     a. to respect the Constitution of the Republic
     b. to respect the Law and Order and promptly fulfil other legal prescriptions
     c. to declare his residence
     d. to provide the authorities with information concerning any alterations in his personal situation or whenever these are asked for by competent authorities
  3. The right expressed in 1 does not include voting rights or other rights and duties reserved by law for national citizens.

On the other hand, Article 14 of the Civil Code provides that 'foreigners shall have the same rights as [Mozambican] nationals with regard to the enjoyment of civil rights, except as otherwise provided for by law'.

Article 20 of Act No. 5/93 stipulates that authorisation for residence shall be granted by the competent government services to foreign citizens holding a residence visa for the exercise of professional activities.

- **Labour Act**
  Article 46 of the Labour Act provides as follows: 'All national or foreign workers, irrespective of sex, race, colour, religion, political or ideological conviction, ascendance or origin, have the right to receive a salary and to enjoy equal pay for equal work'.

7.4 Migration Institutions and Governance

The main Mozambican agency in charge of handling migration issues is the Ministry of Foreign Affairs and Cooperation (MINEC). The subordinate and supervised institution of the MINEC in charge of dealing with migration is the National Refugee Support Institute (INAR). The main responsibilities of the INAR Provincial Delegations are to welcome, protect and give humanitarian assistance to refugees and asylum seekers. Aside from the MINEC, there are a number of other governmental ministries and institutions dealing with migrants and migration matters, including the Ministries of Interior (MINT), Labour (MITRAB), Health (MISAU), State Administration (MAE), and the General Prosecutor's Office (PGR), and coordination entities including the National AIDS Council (CNCS), the National Institute for Disaster Management (INGC) and various municipalities and provincial offices.

International organisations working on initiatives related to migrants in Mozambique involve the UNHCR, IOM Mozambique, UNICEF, UNV Mozambique, UN Women and Pathfinder International.

7.4.1 United Nations High Commissioner for Refugees (UNHCR)

The UNHCR provides comprehensive and solutions-oriented support, including livelihoods and self-reliance activities, to people of concern while advocating for the inclusion of refugees and stateless in the national system. Since August 2020, when the UNHCR declared a level-2 Emergency for Mozambique, the operation scaled up its capacity to respond to the additional humanitarian needs resulting from escalating violence in the northern provinces of the country. Under the clusters approach, the UNHCR leads the protection cluster for the IDPs response on both north and central regions. Finally, the UNHCR adapted its operations to the context of the COVID-19 pandemic to stay and deliver protection and solutions to refugees and IDPs.

In Mozambique, the UNHCR works closely with key government authorities including the National Institute for Refugee Assistance (INAR), Ministry of Health, Ministry of Education and Human Development, Ministry of Gender, Children and Social Affairs (DPGCAS) and the Ministry of Interior. The UNHCR is also working actively with UN agencies and international and local NGOs to support effective delivery of protection interventions to refugees and IDPs in a timely manner, both in refugee camps and urban refugee-hosting locations, and in IDP sites and communities affected by conflict and natural disasters. The UNHCR's implementing partners include Ayuda en Accion, Caritas, Doctors with Africa CUAMM, the Catholic University in Mozambique, the Episcopal Commission for Migrants, and Refugees and IDPs (CEMIRDE), and KULIMA.

The UNHCR, as part of the inter-agency cluster coordination group (ICCG), leads the Protection Cluster – including Gender-Based Violence (GBV) and Child Protection Areas of Responsibility (AoR) – and extensively engages with the Shelter/Non-Food Items (NFI) and Camp Coordination and Camp Management (CCCM) Clusters. In this role, the UNHCR coordinates protection operations for refugees, asylum seekers and internally displaced people (IDPs) in close collaboration with the relevant government institutions, UN agencies and partners. The UNHCR is also leading the network on Prevention of Sexual Exploitation and Abuse (PSEA) jointly with Save the Children, as well as the Community Engagement and Accountability to Affected Population (CE/AAP) Working Group in Cabo Delgado with UNICEF.

The UNHCR is working with multiple stakeholders in Mozambique to promote livelihoods opportunities to refugees and asylum seekers, including government entities, civil society, UN agencies and development actors. More than 800 people from the refugee and local communities have benefited from the second cohort of the Graduation Approach livelihoods programme in the Maratane camp. Based on voluntary choice, participants joined training sessions with specialised private sector companies focusing on tomato, poultry and eggs production, waste management and bio char production, sewing, among others. The UNHCR is also working with MyBucks (MBC) in Nampula to provide bank services in the Maratane settlement to promote the economic inclusion of refugees and asylum seekers. So far, 96% of the Graduation Approach participants completed the opening of bank accounts. Lastly, to promote self-reliance and food security of refugees and asylum seekers, the government has allocated 2,000 hectares of agricultural land for farming purposes.

The UNHCR is strictly implementing COVID-19 preventive measures during humanitarian activities such as focus group discussions with refugees and Non-Food Items (NFIs) distributions by wearing masks and respecting social distancing. In April 2021, the UNHCR donated around 3,500 COVID-19 preventive items for refugees such as goggles, nasal cannulas and disposable gowns to the Provincial Health Service in Nampula. In the Maratane refugee settlement, the UNHCR financed the construction of one isolation and treatment centre and one heath facility.

The UNHCR continues to support the government's efforts through outreach and legal counselling to urban refugees in Nampula and Maputo Provinces. The UNHCR, as the protection lead agency, is also assisting the government in the domestication of the Kampala Convention and setting an example of refugees' inclusion in national planning at all levels. Since 2017, the UNHCR and the INAR are conducting the biometric registration of refugees and the issuance of identification documents. Until the present date,

24,320 ID cards have been provided to refugees living in Mozambique. In addition, the government has been supporting refugees and asylum seekers by issuing declarations stating their right to work NAR.[6]

NAR is responsible for safeguarding refugees' basic needs by assisting with food, shelter, household items, transportation, protection, health services and basic education. It is also responsible for issuing documentation for refugees.

7.4.2 The International Organization for Migration (IOM)

The International Organisation for Migration (IOM) Mozambique runs a broad variety of programmes, including emergency response, refugee resettlement, repatriation and family reunification, migration and development and counter-trafficking. Other IOM activities include the promotion of the international migration law, policy debate and guidance, the protection of migrants' rights, migration health and the gender dimension of migration.

The International Organization for Migration office in Mozambique supports the Government of Mozambique (GoM), partners and national and international organisations to effectively manage migration and to assist vulnerable migrants, displaced persons and affected communities in Mozambique. IOM Mozambique works in the areas of Migration Health, Immigration and Border Management, Migrant Protection and Assistance, Labour Migration and Migration, Environment and Climate Change, as well as other areas.

The IOM's work in strengthening the Mozambican government's capacity in immigration and border management will build on the existing programme with the Directorate of Migration, under which four border points have been revitalised, border officials trained and border management equipment installed. The success of this project has led to an official government request to replicate this project in additional border points and to continue the specialised training for border guards.

The IOM's work concentrates on: (a) providing direct protection assistance to victims of trafficking and stranded migrants, and (b) building protection-sensitive systems and processes for the management of mixed migration.

The trafficking component focuses on three activity areas: (a) capacity-building of prosecutors and other law enforcement officers to understand human trafficking, especially in terms of anti-trafficking legislation in Mozambique; (b) strengthening the quality of services and direct assistance provided; and (c) community empowerment to prevent and respond to human trafficking. The second area focuses on responding to increasing mixed migration flows, including irregular migration of Mozambicans to Southern Africa. This includes developing a national plan of action, establishing protection-sensitive reception mechanisms, revision of legislation and policy for the protection of mixed migrants, building migrants' understanding of their rights, and establishing and supporting domestic and regional coordination mechanisms.

The IOM's work focuses on the following areas: (a) assisting the Government of Mozambique to build its diaspora engagement programme; (b) implementing labour migration programmes with other Lusophone countries and countries in the global South, including protection programmes for cross-border labour migrants; and (c) mainstreaming migration into governance systems, for example, national poverty reduction strategies and plans. The first area focuses on implementing the diaspora engagement strategy, developed in 2014 within the Ministry of Foreign Affairs and in partnership with the Insituto Nacional para as Comunidades Moçambicanos no Exterior, and includes mapping of the global diaspora and maintaining the diaspora website and database. The second area focuses on promoting protection-sensitive labour migration in the context of the Community of Portuguese-Speaking Countries and South–South markets. This includes work that the IOM undertakes with cross-border migrant associations, such as the Mozambican Mineworkers' Association (AMIMO), to build their institutional and technical capacity in advocacy and human rights law.

7.4.3 World Relief (WR)

This organisation has several programs targeting refugees. It has always been involved in activities for refugees in the Maratane refugee camp: agriculture, chicken breeding and micro finance managed by the Community Credit Fund (FCC). In Maputo, the FCC has a micro finance program with 67 refugee clients, while in Maratane camp it has 254 refugee clients, predominantly from Congo, Rwanda and Burundi, following the general refugees' nationality trends in Mozambique.

7.5 Summary

The location of Maputo, on the eastern coastline of the Indian Ocean, makes the city a favourable destination for immigrants who foresee the city as a gateway to opportunities within and abroad. Despite having such a favourable location, the country had witnessed a lot of political instability which, to a large extent, impacted negatively on its migration patterns. The situation has been aggravated by its language barrier – Portuguese – which is the official language, given its political history. However, relative peace being experienced is attracting a lot of immigrants who see a lot of business opportunities in the country. Its favourable migration policies are also helping to lure people to the country despite its poor economic profile.

Notes

1 http://www.ine.gov.mz/estatisticas/publicacoes/anuario/cidade-de-maputo/
2 https://assets.publishing.service.gov.uk/media/57a08aea40f0b652dd0009a2/60828_Youth-vulnerabilties-eco-crisis-Mozambique.pdf
3 Forced Migration Refugee Studies Programme, 2003: 4.

4 *Mozambique: Act No. 21/91 of 31 December 1991 (Refugee Act)* [], 13 December 1991, available at: https://www.refworld.org/docid/3ae6b4f62c.html [accessed 27 July 2021].
5 *Mozambique: Law No. 5/93 of 1993 (aliens)* [Mozambique], 28 December 1993, available at: https://www.refworld.org/docid/4485a0f04.html [accessed 26 July 2021].
6 NAR stands for Núcleo de Apoio aos Refugiados.

Bibliography

De Sousa Santos, B. 2018. *The End of the Cognitive Empire: The Coming of Age of Epistemologies of the South*. Duke University Press.
Instituto Nacional de Estatísticas. 2019. *Anuario estatístico*. Maputo Cidade.
Isen, J. & Halperin, L. 2003. Report on FCC/WR micro credit programs to war refugees in Mozambique. Cited in Malauene, M. M. & Landau, L. 2004. *The impact of the Congolese forced migrants' 'permanent transit' condition on their relations with Mozambique and its people* [University of the Witwatersrand].
Jacobsen, K. 2001. *The Forgotten Solution: Local Integration for Refugees in Developing Countries*. UNHCR.
Jacobsen, K. 2004. Microfinance in Protracted Refugee Situations: Lessons from the Alchemy Project. A Paper Presented in Alchemy Workshop held in Maputo, Mozambique. February.
Kibreab, G. 1999. Revisiting the Debate on People, Place, Identity and Displacement. *Journal of Refugee Studies*, 12, 384.
Lloyd-Jones, T. & Rakodi, C. 2014. *Urban Livelihoods: A People-Centred Approach to Reducing Poverty*. Routledge.
Majale, M. 2001. Towards Pro-Poor Regulatory Guidelines for Urban Upgrading. A Review of Papers Presented at the International Workshop on Regulatory Guidelines for Urban Upgrading held at Bourton-on-Dunsmore, 17–18.
Malauene, M. M. & Landau, L. 2004. *The Impact of the Congolese Forced Migrants Permanent Transit'condition on their Relations with Mozambique and its People*. University of the Witwatersrand.
Miamidian, E. & Jacobsen, K. 2004. Livelihood Interventions for Urban Refugees. Paper Written for Alchemy Project Workshop, Maputo.
Migrants & Refugees. 2021. *Migration Profile, Mozambique*. Vatican City: Migrants & Refugees Section | Integral Human Development. https://migrants-refugees.va/country-profile/mozambique/
ODI and CESC – Civil Society Learning and Capacity Building Centre. 2011. Youth vulnerabilities to economic shocks: A case study of the social impact of the global economic crisis on youth in four neighbourhoods in Maputo City, Mozambique.
UNHCR. 2020. Zimbabwe. *Fact Sheet*. https://reporting.unhcr.org/sites/default/files/UNHCR%20Mozambique%20fact%20sheet%20December%202020.pdf

8 eGoli

Beyond the Splendour of Johannesburg through the Eyes of Migrants

8.1 Johannesburg, South Africa

Johannesburg has seen waves of different peoples occupying the area; these include the Stone Age ancestors dating back 500,000 years; the Khoi and San from 1,000 years ago; the 500-year-old Iron Age furnaces belonging to Tswana people; and Boer farmhouses dating from the 1860s. But present-day Johannesburg, 'the city of gold', owes its existence to the discovery of gold on the Witwatersrand (Shorten, 1966) by Australian gold prospector George Harrison. At the time, the Witwatersrand was within the borders of the Boer Republic of the Transvaal (Tomlinson et al., 2003; Abrahams et al., 2018). Discovered in 1886, the main reef turned out to be the largest and richest goldfield ever discovered in the world (Mandy, 1984; Robertson, 1986). This incident set in motion one of the most rapid urban expansions recorded within the 20th century (Chipkin, 1993). Officially, the city was born by a proclamation that appeared in a Government Gazette on 8 September 1886 and was read to a gathering of gold diggers on the spot of the discovery, with such pomp and ceremony (Symonds, 1953). By then, the city-to-be was only a collection of ox-wagons and tents known as Ferreira's Camp (Symonds, 1953). In 1897, Johannesburg acquired a republican-style municipal system (Maki, 2010). Today, the man-made mountains of mine dumps, narrow streets and high-rise buildings of modern and diverse architecture are constant reminders of the city's turbulent origins as a mining town and serve as birthmarks from which Johannesburg cannot escape (Mandy, 1984).

The growth and expansion of Johannesburg was phenomenal from the onset. As the news of gold discovery spread across the Transvaal – from the diamond fields of Kimberly in the west, to the gold mines of Barberton in the East – steady trickles of diggers and prospectors soon combined to create a flood the 'poured with a swirling rush' into what is now Johannesburg (Symonds, 1953; Mandy, 1984). Within a short space of time, the news of the discovery acted as a magnet not only to prospectors already scattered throughout the Transvaal, but also to those from across the world who were drawn by the prospect of riches (Robertson, 1986). Within a decade of its establishment, Johannesburg had become a dominant economic force and the largest city in South Africa, and within 60 years, it became a metropolis

DOI: 10.4324/9781003184508-8

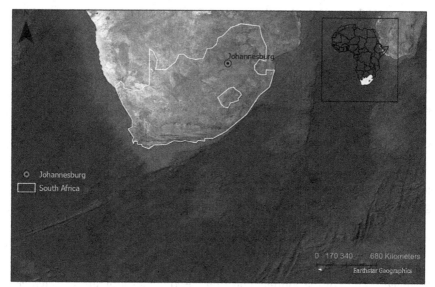

L. Chipungu

Figure 8.1 Map of South Africa Showing the Position of Johannesburg

of unprecedented modernity (Knox and Gutsche, 1947; Abrahams and Everatt, 2019). From 1886, it has been rebuilt four times – starting out as a tented Mining Camp, it evolved into a Town of tin 'shanties', followed by four-storey Edwardian brick buildings and then a city of modern skyscrapers.

In 1889, Johannesburg had only 15,000 inhabitants and by the time the first census was held in 1895, the population had risen to over 80,000, of which more than half was white. In just nine years of its establishment, Johannesburg had become the largest city in South Africa, far surpassing Cape Town, which was already in existence for more than 200 years. By 1936, it was recognised as the largest and most densely populated 'European City in Africa' (Murray, 2011). Today, Johannesburg is the largest city in Southern Africa and is the region's leading industrial and commercial centre (Abrahams et al., 2018).

As will be demonstrated, present-day Johannesburg continues to attract people who are constantly searching for better life, a place to settle, live, learn and work (Abrahams and Everatt, 2019). On a daily basis, the city draws in new arrivals (both short and long term) from across the continent and beyond, setting in motion a process of urbanisation that is driven largely by economic opportunities that the city offers (Turok, 2012; Oosthuizen and Naidoo, 2004).

After the establishment of democratic rule in 1994, Johannesburg again asserted itself as a globally linked economic powerhouse with a trade and investment reach spanning across Africa and beyond (Murray, 2011).

Currently, Johannesburg has the most advanced economy in Africa endowed with power supplies and a well-knitted infrastructure network developed during last 25–30 years (Parilla and Trujillo, 2015). The current size of Johannesburg's economy ranks in the same group as many of the cities of the Middle East whose growth rates are significantly high (Oxford Economics, 2016). On a global scale, Johannesburg ranked as the most competitive city in Africa in 2013 and was ranked 66th in world (*The Economist*, 2013). The city was also named the 33rd most economically powerful city globally, out of 84 leading Financial Cities in the 2015 edition of the Global Financial Centres Index.

While the extractive industry was prominent in the early stages, this shifted towards the secondary sector in the 20th century, combining to form a conglomerate of economic activity (Abrahams and Everatt, 2019). The current economic fabric of Johannesburg is dominated by tertiary and secondary economic activities such as trade, investment, finance, tourism, accommodation, telecommunication, manufacturing and educational services. The city also holds the most wealth on the African continent at $248 billion; it is the financial capital for the continent hosting the Johannesburg Stock Exchange, the largest stock exchange on the continent and the 16th biggest in the world. It is therefore not very surprising that more than 70% of South Africa's companies and headquarters of major corporates, both local and foreign, are located in Johannesburg. Johannesburg generates 16.5% of the country's wealth and employs 12% of the national workforce. Many migrant job seekers and economic immigrants choose Johannesburg as their priority destination.

Centrally located within Gauteng province, together with City of Tshwane (formerly Pretoria), Ekurhuleni Metropolitan Municipality and the West Rand, Johannesburg currently forms part of a large polycentric city-region called Gauteng. The city-region, which has Johannesburg as its centre and key socio-economic driver, reflects the challenges and opportunities of South Africa's extraordinary economic, demographic, social and political transformation (Parilla and Trujillo, 2015).

It stretches for approximately 80 km in length and 50 km in width along a northeast/south-west bearing (Storie, 2014). It is situated in the interior Highveld grassland plateau of South Africa, and stands at an altitude of 1,740 m above the sea level (Fatti and Vogel, 2011). The city of is one of the 10 local municipalities that make up Gauteng Province. Because of its centrality, Johannesburg is a major transportation hub linking major ports and trade partners in the region such as Zimbabwe, Mozambique, Botswana, Malawi and Zambia.

With more than a century of residential suburban development, Johannesburg has developed a unique ecological profile and has the world's largest urban forest with an estimated 10 million trees (Schäffler et al., 2013). This urban forest consists of a variety of flora and fauna species and has a subtropical climate. The average midday temperatures range between 19°C to 7°C (Schäffler et al., 2013). However, the city's environment is under severe threat from a number of factors associated with a rising population, densification and urban sprawl (SACN, 2018), particularly in the CBD area, township areas and informal settlements.

Johannesburg began as a gold mining town, attracting investors, opportunists and workers from the early 1900s. According to scholars, the city became a site of unprecedented modernity within a few decades of its establishment (Knox & Gutsche, 1947; Mandy, 1984). It has been called 'A World Class African City', the City of Gold, and the City of Millions of Lights. Yet it also has the odious title of being the most unequal city in South Africa, and the world, according to various reports. Present-day Johannesburg continues to attract new people who are searching for a better life, a place to settle, live, learn and work (CoJ, 2011), driven largely by economic opportunities that the city offers (Holland and Roberts, 2002; StatsSA, 2015).

The metropolitan municipality of Johannesburg (hereafter Johannesburg) as it is known today represents one of the most diverse cities in the African continent. Its urban cosmopolitanism, however, is marred by appalling inequality and this strongly relates to the spatial form of the city. Apartheid spatial planning has meant that Johannesburg has been ranked as the most unequal city in the world, in the most unequal country in the World (Euromonitor, 2017).

8.2 Migrants Profile in Johannesburg

Johannesburg is the main economic hub in the region, attracting migrants both from within South Africa and from around the continent. The post-1994 period saw new opportunities emerging for immigrants as South Africa reconnected with the global economy (Crush et al., 2005). The result was an increase in the flow of people from other countries in the region and the world beyond (StatsSA, 2015). Compared to the pre-1994 period, the pattern of international migration shifted as highly skilled and low-skilled immigrants, refugees, asylum seekers and illegal immigrants became more prominent (StatsSA, 2015). The major pull factors include the availability of social infrastructure, educational opportunities and medical infrastructure, contrasted with political unrest and economic decline in the neighbouring countries (Cohen, 2008). In 2013, the City of Johannesburg had the highest proportion of migrants in the 20–24 to 30–34 age cohorts, totalling 45% (Parilla and Trujillo, 2015). The Gauteng Province, where Johannesburg is located, experienced the highest rate of growth in population between 2001 and 2011 at 30%; half (52%) of that was a result of immigration (Peberdy, 2013).

The city is home to both local and foreign migrants. The local migrants come from all corners of the country in search of a better life through employment or educational or business opportunities. In terms of the cross-border migrants, it includes those that have come into the country through work and study permits, tourists, refugees, asylum seekers, holiday visa over-stayers and undocumented/irregular migrants. South Africa hosts approximately 4,036,696 international migrants, of which the male population constitutes 2,244,421, while the female population constitutes 1,792,275. Children constitute 15.90% and the elderly constitute 5.60% of the migrants. Around 7.66% of migrants consist of refugee and asylum seekers (UN, 2017).

According to a study conducted by the Centre for Development and Enterprise (CDE) in 2008 the Zimbabweans and Mozambicans were seen by Johannesburg residents to be the largest groups, followed by Nigerians, Chinese and Malawians; at a lower level were Indians, Pakistanis and Europeans. Immigrants from other countries in Africa have low visibility because of their small numbers, although Somalis have low numbers but high visibility. Those from Botswana, Lesotho, Namibia and Swaziland probably exceed Nigerians but have low visibility. Foreigners are perceived to work as hawkers, artisans, miners, domestics and gardeners, shopkeepers and stall-holders, in professional and technical activities, personal services, security, and small manufacturing and crafts. There is, however, lack of accurate statistics regarding the migrant population in the city due to a number of factors, including the difficulty to document irregular migrants who avoid being identified and difficulty of distinguishing between migrants who have settled here and those who regularly return home.

Categories of migrant living in Johannesburg vary from those who are temporary migrants and those who are long-term residents, family members and marriage partners of established immigrants. Temporary migrants can be regarded as those that are in the city for a specific period as determined by their visa conditions and includes all those on legitimate study, holiday and work visas. In as far as the local migrants are concerned those that can be considered to be temporary are those who have come to undertake a specific task and then return back to their homes.

Migration has resulted in an increase in diversity and a multicultural society, both of which are the foremost features of globalisation. Because of these differences and host country's socio-economic dynamics, migrants face numerous difficulties in their countries of destinations. South Africa's constitution provides protection for everyone regardless of their legal status or nationality (Okyere, 2018). Notwithstanding that, migrants, especially refugees, continue to experience challenges in their daily lives (Smit and Rugunanan, 2014). These challenges include segregation from the labour market, lack of access to health services, housing and welfare, as well as social exclusion (Datta et al., 2006).

The interviews contacted with government officials and academic professionals revealed that migrants face enormous challenges in accessing adequate housing. The question of housing in South Africa has been a thorny and urgent issue for decades. The basic needs of migrants and refugees must be met for them to integrate successfully into society and one of the most significant, basic needs is affordable, suitable and adequate housing (Teixeira and Halliday, 2010). The South African constitution also stipulates that everyone has the right of access to adequate housing, with the South African Refugee Act (130 of 1998) guaranteeing those rights provided in the constitution to migrants, particularly refugees (Greenburg and Polzer, 2008). With many South Africans living in shacks and overcrowded spaces in townships and informal settlements, the situation of migrants appears to be even more precarious and difficult. Most migrants in South Africa are ordinary hard-working people, who do not depend on state support for housing. Migrants, just like South Africans, have a variety of housing options in urban centres. There is the private sector housing, which is by far the most significant sector for migrant housing, followed by informal accommodation and public housing (Greenburg & Polzer, 2008). However, though some can afford to pay for adequate housing, their uncertain legal status often leaves them living in totally inadequate housing and struggling to access the 'adequate' housing that the Constitution guarantees.[1]

Neighbourhoods in which houses are found and the types of houses they inhabit influence migrants' access to social networks, employment opportunities and their general sense of security (Teixeira and Halliday, 2010). In Johannesburg, migrants settle and work in the inner city and in the townships on the outskirts of the city. Those in the inner city have better access to markets, housing and public services. In contrast, the townships, or 'locations', are characterised by poor physical infrastructure, inadequate education and poor health outcomes. Compared to residents in the inner city, residents in the townships are more likely to be unemployed, live in poor housing conditions, have minimal access to services and be more vulnerable to violence (Monson et al., 2010). In this unstable environment, forced migrants, economic migrants, internal migrants and poor South Africans compete for limited urban resources. Crime in Johannesburg and the townships is high. Poor neighbourhoods are characterised by high incidences of violent crime, such as rape and homicide (Amit, 2010). Threats of violence are often made against those who are already the most socioeconomically vulnerable and those perceived to be stealing jobs (Monson et al., 2010).

Because of the difficulties that migrants have in accessing housing legally, many South Africans offer them housing at much higher prices (Peberdy, 2016); others refuse to give back their deposits at the end of the lease, and many other forms of exploitation of migrants take place. This exploitation exacerbates the conditions of already poor migrants. Poverty also means that people end up living in shacks, abandoned buildings or other dangerous spaces because they cannot afford to live anywhere decent. Poverty also means many are forced to find places closer to work, and these are often expensive or overcrowded.

Given all these problems, the option of accessing social housing becomes close to impossible, making migrants even more vulnerable and poor. Banks do not readily give home loans to migrants due to their lack of stable domicile. The result is that the major source of housing for migrants becomes private rental housing, and this has its own challenges. Volatility is another key characteristic of housing histories: many migrants and refugees describe histories of continuous displacement from one temporary form of accommodation to the next (Greenburg and Polzer, 2008).

Migrants are excluded from public housing systems in South Africa; housing subsidies are restricted to citizens and the country's permanent residents. Categories of legal migrants with rights that are nearly the same as those enjoyed by South African citizens, such as asylum seekers and refugees, are also excluded from the public housing subsidy. Despite these exclusions, there are policymakers who have recognised legal migrants' right to housing schemes in South Africa. However, they attribute the lack of implementation of rights-based access to public housing to technical and systematic problems with documentation procedures.

Non-governmental actors often provide limited shelters to the migrant population; this form of accommodation is, however, temporary and lacks funding. In addition to this, non-citizens from the inner cities are often excluded from these shelters, because of budgetary constraints, and some are also denied access because these shelters claim they have no rights to space. As a housing (http://etd.uwc.ac.za/) survival strategy, migrants take up long and short-term accommodation in churches, which have proved to be a significant provider of housing needs for this population (Greenburg and Polzer, 2008).

All these issues persist and make the housing condition of migrants extremely unstable and definitely inadequate, falling far short of what is required by the Constitution. Housing does not only provide physical shelter but strongly impacts the type of livelihood and health options (Greenburg and Polzer, 2008). Challenges faced by migrants in accessing health care indicated a high score, which reveals the vulnerable conditions migrants find themselves in Johannesburg.

The health of migrants, just like that of non-migrants, is influenced by biological factors, socioeconomic status, behaviour and exposure to the environment. In addition, migrants' health status may be influenced by the health risk profile that characterises their country of origin or may arise in the migration process (Gushulak and MacPherson, 2006). That said, the migration process can also influence healthcare services in transit and destination countries and a high demand may be placed on the services in these countries as a result (MacPherson et al., 2007). This high demand may be due to the number of migrants or due to migrants having different diseases in comparison to the host country (MacPherson et al., 2007).

Vearey and Nunez (2010) in their research, found that cross-border migrants in South Africa are denied access to healthcare services because of the prevailing discourse in the healthcare system, which associates health seeking as one of the main drivers of international migration. In this particular instance, healthcare service providers were found to ration services to migrants to limit migration. Furthermore, health caregivers encounter various challenges in managing and caring for migrants. These challenges include language and culture barriers, and resource constraints of migrants and inadequate health systems (Suphanchaimat et al., 2015). Undocumented migrants, because of fear of deportation, may not seek health care services whenever they need to do so. Buttressed by Hacker et al. (2015), migrants have reported avoiding health facilities and waiting until their health conditions become serious to seek services because of their worries about being deported. The exclusion of migrants, often from social protection policies such as pensions, unemployment benefits, and safety-net programmes like food transfers and health insurance, results in marginalisation and social insecurity (WHO, 2010). These elements impact the health of migrants and their productive integration into society negatively (WHO, 2010).

The basic human right of access to health services is incorporated in the constitution of South Africa and the progressive realisation of this right is being recognised despite limited resources (Walls et al., 2016). Consequently, free primary health care at the point of use is included in South Africa's public health system. This includes free health care for all lactating and pregnant women as well as children under six years of age and free primary and emergency health care at the point of use for all (Walls et al., 2016).

Crush and Tawodzera (2011a) found that migrants seeking public health care in South Africa are required to provide their identity documentation, evidence of residence and proof of a home address before treatment is provided, effectively excluding all migrants who are undocumented or without formal housing or a formal home address. Migrants in need of antiretroviral therapy for HIV have also been found to encounter challenges in the public health institutions. Thus, they are compelled to rely on the NGO sector for antiretroviral therapy (Crush and Tawodzera, 2011a). This discrimination towards migrants in accessing public health care in South Africa is termed 'medical xenophobia' (Crush and Tawodzera, 2011a). Medical xenophobia is the negative attitude that health workers exhibit towards refugees and migrants (Crush and Tawodzera, 2011a). As a survival strategy, these migrants resort to buying medicines at local pharmacies and visiting traditional healers for health care. Some resort to their social networks by asking acquaintances who have documentation to access non-prescription drugs over the counter or get others to visit the clinic, pretend to be ill and thus access medicines on their behalf (Crush and Tawodzera, 2011a).

Integration is a social process that occurs when native people settle or coexist with immigrants (Ogbu and Matute-Bianchi, 1986). In their new communities immigrants often experience initial problems of adjustment in the schools, although their problems are not necessarily characterised by persistent adjustment difficulties or low academic performance (Ogbu and Matute-Bianchi, 1986). While immigration is not a phenomenon unique to South Africa, certain conditions under which immigrant learners find themselves are typical of the South African context. These conditions, as presented in local studies (Mbhele, 2016; Vandeyar, 2010), include academic and social exclusion owing to a lack of proficiency in indigenous languages, moral degeneration among indigenous learners and conflicting South African–African values, stereotyping and xenophobia, unaffordable schooling (owing to costly transport and uniform). These conditions imply that while immigrant learners can access schools, as the law gives them the right to do so, they face challenges in classrooms and on school grounds, which constitute barriers to their education. But of critical importance is that the right to education for these children is broader than their access to schooling, as one may be tempted to think. This right extends beyond access to cover quality education (Craissati et al., 2007).

There is a common assumption in South Africa that these children and students have no right to an education in South Africa. Researchers and NGOs have already pointed to some of the problems that migrant children and their parents face in accessing government schools in South Africa. These include demands by school administrators and principals for study permits and birth certificates, language admission tests, claims that schools are 'full', being relegated to the bottom of enrolment lists, financial hardship, geographical inaccessibility and unwarranted fee demands (Crush and Tawodzera, 2011b). Given the extra guarantees of the Refugees Act, child refugees should have no difficulty at all in gaining access to schools. But parents who have come to South Africa to seek asylum are frustrated by South Africa's overburdened and ineffectual refugee system. As Crush and Tawodzera (2011b) point out, the refugee determination process can take months, if not years, leaving asylum seekers in 'a state of limbo' during which they may stay in the country, but can access few social services and receive almost no official or private assistance in the form of direct aid.

Hirschman (1982) found that human capital, which includes work experience that can be useful in the labour market and educational skills, is one of the important prerequisites for migrants' socio-economic integration in their host countries. Even though migration, the world over, is mainly triggered by economic factors, one of the challenges faced by most migrants, nonetheless, is accessing a livelihood. South Africa's constitution provides protection for everyone regardless of their legal status or nationality. Notwithstanding that, migrants, especially refugees, continue to experience difficulties in their daily lives (Smit and Rugunanan, 2014).

Landau and Jacobsen (2004) note in their study of a sample of refugees in Johannesburg, South Africa, that these refugees experience myriad constraints in trying to access their livelihood, which includes a tiresome procedure of lodging an asylum claim, difficulty in gaining legal documentation from the Department of Home Affairs which will provide them refugee status and harassment from the police. Because of their readiness to work for a low wage and under unfavourable conditions, many refugees remain in a weak economic condition and many resort to working in the informal sector (Landau and Polzer, 2008). Moreover, Northcote (2015) found that most migrants, whether forced or legally recognised, have trouble in finding employment in the formal economy in South Africa. This is identified as one of the most anxiety-provoking challenges for refugees and asylum seekers (Smit and Rugunanan, 2014). Since most migrants in South Africa are unable to access employment in the formal sector, many resort to the informal sector for a source of livelihood (Northcote, 2015). The difficulty of finding formal employment for most migrants is linked to challenges in accessing proper documentation. Many are unable to access the country's Department of Home Affairs for a legal migrant status, have difficulty in accessing the banking systems and are marginalised by a perception among citizens that foreign migrants steal jobs (Northcote, 2015). Thus, migrants who are largely excluded from the formal sector exhibit high levels of creativity and enterprise in the informal market (Crush et al., 2015).

Migrants continue to play an important role by providing labour (skilled and unskilled) and others contribute to the economy through cross-border trading (Oosthuizen and Naidoo, 2004; Peberdy, 2013). In the same vein, a number of popular and academic writers have reported that African immigrants in Johannesburg have not been treated well. The interviews conducted with academic professionals reveal that access migrants face hardship in accessing employment and other economic opportunities. This explains why they are limited to informal economic activities. True to 'the entrenched and systemic' xenophobia 'in South African society' (Dodson, 2010), violent attacks on and deaths of African immigrant traders by South Africans have been recorded. For example, Landau and Polzer (2007) chronicled detailed incidents of the attacks, some of them fatal, against African foreign nationals in general and informal traders specifically. Johannesburg in 2015 saw widespread attacks on international migrant businesses, mostly in the informal sector. Although international migrant businesses and their owners have been for many years, and still are, the targets of xenophobia in South Africa, the violence of 2015 led to renewed public debate about their place in South Africa. This is evidenced by the attacks on Somali shops by the residents of Ramaphosa settlement in June 2011 (*The Star*, 2 June 2011, p. 6) and in May 2013 (*The Star*, 29 May 2013, p. 4). A newspaper (*The Star*, 19 May 2011, p. 6) reported that the harassment of foreign traders in Reiger Park (Erkurhuleni Municipality), in Soweto and in other townships around Johannesburg was led by a formal business forum. The incidents of xenophobia in the latter two references show that foreign shop owners are vulnerable victims who cannot return to their countries of origin as the reasons that motivated them to leave their countries of origin and come to South Africa still persist. In January 2015, there was an attack on foreign owned businesses, especially those owned by the Somalis in Johannesburg (Hlubi, 2015). In April 2015, foreign-owned shops and especially those of African immigrants were looted in Jeppestown, Johannesburg, in what is considered as a xenophobic rage (De Klerk, 2015; Hawker, 2015) and similar incidents were reported in Alexandra, Johannesburg (Aboobaker, 2015).

Issues of migrant integration is influenced by the state of local economies, geographic location, history and culture, are all factors responsible for shaping the context of reception for new migrants (Martinez et al., 2012); furthermore it's influenced by government policy and regulation, as well as institutions as detailed in the section below.

8.3 Migration Policies and Regulatory Environment

South Africa's current immigration policy makes a clear statement in favour of skilled migration. In the African context, South Africa's immigration system is quite sophisticated and offers a lot of options for immigrants; the challenges are on the implementation side. The very recent Constitution of South Africa is widely regarded as one of the most progressive in the world. This has direct impact on the immigration framework.

Due to the complexity surrounding the question of immigration in South Africa, which is characterised by a diversified immigrant population whose purpose and motivation for seeking South Africa as a destination country are also varied, it is logical that the different categories of immigrants are not regulated by a single umbrella legislation. The Constitution alone may be said to have an umbrella effect because it provides the statutory framework for every other national legislation such as the Refugee and the Immigration Acts. The primary immigration-related statutes are the Immigration Act, the Refugee Act and the Constitution. The existing policy on international migration was set out in the 1999 White Paper on International Migration. It was implemented through the Immigration Act and partly through the Refugee Act.

8.3.1 Constitution of the Republic of South Africa 1996

The South African Constitution is the overarching law of the land to which every other law and all conduct, including by the government and organs of state must at all times be consistent with (Republic of South Africa, 1996). The obligations that it imposes must therefore be complied with (Republic of South Africa, 1996). The Constitution does not make provision for issues relating directly to immigration but lays down general principles of law and standards of behaviour that form the basis of a democratic an open society established on democratic values, social justice and fundamental human rights (Republic of South Africa, 1996). It guarantees the fundamental human and socio-economic right to equality under which immigrants who are legitimately within the country may seek protection. The equality clause states that:

1. Everyone is equal before the law and has the right to equal protection and benefit of the law.
2. Equality includes the full and equal enjoyment of all rights and freedoms. To promote the achievement of equality, legislative and other measures designed to protect or advance persons, or categories of persons, disadvantaged by unfair discrimination may be taken.[2]

The Bill of Rights enshrines, among others, the rights to property, housing, health care, food, water and social security.[3] According to a purposive interpretation of the Bill of Rights as stipulated by art 39(1)(a) of the Constitution, which must aim to 'promote the values that underlie an open and democratic society based on human dignity, equality and freedom', immigrants are not excluded from the constitutional rights. It is noted, however, that 'hardly any South African citizens or politicians see foreigners as entitled to these rights' (Laher, 2008).

8.3.2 Refugee Act 130 of 1988

South Africa is a signatory to the 1951 Refugee Convention, the 1976 Protocol relating to the Status of Refugees and the 1969 Organization of African Unity Convention Governing the Specific Aspects of the Refugee Problem in Africa – agreements that detail the rights of recognised refugees. The Refugee Act of 1988 was adopted as a legislative measure to translate the international commitments undertaken by South Africa under the 1951 UN and the 1969 OAU (AU) Refugee Conventions into domestic realisation. The Act defines the process by which someone can apply for asylum in South Africa (Mthembu-Salter et al., 2014). It grants asylum seekers and refugees' freedom of movement, the right to work and access to basic public services, such as health care and public education. It envisages:

> To give effect within the Republic of South Africa to the relevant international legal instruments, principles and standards relating to refugees; to provide for the reception into South Africa of asylum seekers; to regulate applications for and recognition of refugee status; to provide for the rights and obligations flowing from such status; and to provide for matters connected therewith.[4]

Besides the obligations imposed by the UN and the OAU (AU) Refugee Conventions, the Refugee Act that was drawn up in 1998 and enacted into law in the year 2000 also enjoins the government of South Africa to receive, accommodate and protect persons who have been compelled to leave their countries of origin because of well-founded fear of persecution, violence or conflict.[5] By adopting domestic legislation to complement its international engagements, South Africa reaffirms its commitment to be bound by its own law in providing protection to asylum seekers and refugees within its borders, not only as a humanitarian gesture but essentially as a legal obligation.

The Act allows South Africa the authority to set the principles and standards relating to the reception of asylum seekers; regulate the asylum application process; and the conditions for the granting of refugee status. It also provides, among others, for the rights and obligations that refugees and asylum seekers are legally entitled to. As the National Consortium for Refugee Affairs has outlined, the following rights can generally be deduced from the Refugee Act of 1998:

- They have the right to not be returned to their country of origin or any other country if doing so would place their life or security at risk;
- From the moment they lodge an asylum application, they have the right to work and study. Refugees have access to health care, public relief, and assistance. All people in South Africa have the right to life saving medical treatment;
- The right to have their asylum applications adjudicated in a manner that is lawful, reasonable and procedurally fair, which includes the right to appeal a negative decision on asylum claim;

- The right to freedom of movement and not to be arbitrarily arrested and detained. When detained, under conditions consistent with human dignity; and
- The right to legal representation.[6]

The Act specifically provides that once granted legal status, refugees are entitled to full legal protection and the enjoyment without discrimination of all the rights and privileges set out in the Bill of Rights.[7] Emphasis is placed on the rights to seek employment, access basic health care and education.[8] Thus, the same rights that are guaranteed to all South African are equally guaranteed to refugees for as long as they discharge of their legal responsibility to respect and stay within the confines of the laws of the country.[9]

The Refugee Act prohibits the government from refusing entry into the country, expelling, extraditing or returning to any other country or be subject to any similar measure, if as a result of any of such action the persons concerned are exposed to persecution or threat to life.[10] The legal obligations imposed by the Refugee Act in conjunction with the Refugee Conventions mentioned above entail that South Africa may not forcefully repatriate the asylum seekers and refugees within its borders except for reasons that are genuinely justified by law. The perpetuation of any action by the government or the allowance of such action to be perpetuated by a third party would be deemed to contravene both international as well as domestic law.

South Africa's progressive refugee and asylum legislation with an impressive guarantee of rights is inspiring, probably accounting for the massive influx of foreign nationals to seek refuge in the country.[11] However, it is noted that although these rights are provided for by law, accessing them remains an illusion for many refugees and asylum seekers.[12] In spite of these assurances of protection, the two instances of widespread xenophobic eruptions, which affected large numbers economic immigrants as well as asylum seekers and refugees, hold evidence of the government's indifference in providing sufficient protection as stipulated by the law. The government's indifference has often manifested in the fact that instead of taking decisive measures or concrete action in dealing with the crisis, rather tried to mitigate the gravity by claiming that xenophobic attacks are acts of criminality.

The Refugee Act places the responsibility upon the South African government to provide full protection and provision of rights set out in the Constitution – this includes access to social security and assistance (McConnell, 2009). As of 12 January 1996, South Africa is bound by ratification to the UN Refugee Convention, obliging the state to provide equal treatment to refugees as it would to its nationals. Meanwhile, though the Immigration Act attempts to be more migrant-friendly, it is considered extremely limited and ambiguous, with emphasis almost exclusively focused on attracting highly skilled migrants. It may be argued that such a policy is discriminatory and, therefore, contravenes international human rights law that provides for the right to movement to everyone without discrimination of any form[13] as well as the constitutional law guarantee of equality and non-discrimination (McConnell, 2009).

Compliance with the law entails the government to take its obligation to ensure sufficient protection of the foreign nationals in the country seriously and most importantly to ensure that they make a valuable contribution in transforming the country's socio-economic landscape. On this note, it is essential to look at the legal instrument that regulates the category of immigrants who come to South Africa for other reasons than to seek asylum.

8.3.3 Immigration Act 13 of 2002 as amended in 2014

The South African Immigration Act of 2002, as amended in 2011, to address glaring gaps and mistakes in the legislation, and only promulgated in May 2014, is the primary legal instrument that regulates and sets out conditions of entry, residence and departure of the broader range of immigrants, including temporary visitors of different category, economic migrants, investors and permanent residents. The Act lays down conditions and procedures for obtaining visas and permits for skilled migrants, students, tourists and other categories of permanent and temporary migrants and also envisages the arrest and deportation of undocumented migrants (Polzer, 2010). South Africa's immigration system is not based on a quota or points-based system. It does not distinguish between the right to work and the right to reside. Unlike countries such as Botswana, Mozambique or Tanzania, the work visa or other form of visa also gives the right to reside in South Africa. Despite recommendations to the government to relax its immigration policy, the 2014 amendment of the Immigration Act rather 'increased the barriers to migration for all categories of migrants' (Mthembu-Salter et al., 2014). The most recent amendment to the Immigration Regulations was in November 2018.

The Immigration Act does not provide for the same kind of rights and privileges as guaranteed by the Refugee Act, which means that the government is under no particular human right obligation towards immigrants whose entry into the country is regulated by the Act. However, because the Act is presumed not to be inconsistent with the Constitution that embraces 'all who live' in South Africa, it is logical to argue that the constitutional guarantee of equality and non-discrimination allows foreign nationals who are legitimately within the country on regular visas or permits to enjoy the same rights and to exercise the same responsibilities as South African nationals. Polzer (2010) has ascertained that the Bill of Rights 'grants all people in South Africa – citizens and both documented and undocumented non-citizens – rights to life, dignity, equality before the law, administrative justice, basic education, basic health care, and labour rights' (Polzer (2010)). The Constitution also guarantees to *everyone*, including foreign nationals and their dependents, the right to social security and assistance.[14]

The government is required to structure its immigration policy to be more receptive, accommodating and immigrant-friendly in view of combining the imperative for socio-economic transformation and the need to protect the national interest. However, it seems that the constitutional changes introduced with the coming of the new political dispensation created a framework in which various interest groups are struggling to consolidate their positions and, therefore, see immigrants as a new threat that needs to be dealt with (Kabwe-Segatti, 2008).

Though government thinking on immigration has in principled changed in favour of a policy that takes into consideration the contribution that immigrants can make to the South African society in the largest sense, anti-immigrant sentiments are still very much prevalent. The immigration policy restricts the entry of unskilled migrants into the country and rather encourages 'highly skilled migrants to work in key sectors of the economy. In spite of this, it is noted that significant political constituencies and administrative services remain ambivalent or even opposed' to the idea of an open and receptive immigration policy, making it difficult for immigrants to enter and stay in the country legally.

Since the end of 2015, the Department of Home Affairs (DHA) has started a comprehensive review of the existing international migration policy. Essentially, the country's formal international immigration policy had remained in place since 1999 despite significant changes in the country, region and in the world. The major problem of the existing policy is its inability to enable South Africa to adequately embrace global opportunities while safeguarding its sovereignty and ensuring public safety and national security. South Africa has not yet built consensus on how to manage international migration for development. There is a lack of a holistic and whole-of-government-and-society approach. The existing policy is based on an approach that is largely static and is limited to compliance rather than to managing international migration strategically to achieve national goals. In addition, there is no sense of South Africa being an African state situated in the SADC, which is one of the eight regional communities recognised by the African Union (AU).

The former Minister of Home Affairs, Malusi Gigaba, had therefore identified the development of a new international migration policy as one of the top priorities during his first term in office. As the first step towards this, the Green Paper on Migration was published in June 2016, and the next step in the legislative process was the White Paper on International Migration published in July 2017. The new immigration bill was introduced to parliament during the course of 2020. In addition, the National Development Plan (NDP) makes a clear statement in favour of immigration for the future. The NDP essentially argues that if South Africa is to end poverty and create decent work it must use migration to grow the skills and knowledge base and remove barriers to regional development. This requires South Africa to invest strategically in the further development of an efficient and secure immigration system.

In practice, advocates for forced migrant rights argue that the government does little to guarantee access to these services. South Africa has no coherent legal framework for dealing with the high numbers of forced migrants arriving at its borders. Little attempt has been made to establish a clear and consistent national migration policy that provides for effective migration management systems and structures to ensure that the rights of migrants are protected.

The development of the policy on integrating migrants into the social fabric of Johannesburg society is primarily based on the experience of the Johannesburg City Council's efforts after years of implementing these policies and developing the Counter Xenophobia and Common Citizenship Programme and spearheading intervention to integrate migrants into communities.

8.4 Migration Institutional and Governance

The DHA has historically been regarded as the sole department responsible for managing international migration. As well as immigration tasks, it is also in charge of all civic matters for South African citizens and permanent residence holders, and its staff is deployed at the borders. The adjudication of all immigration visas and permits is done by the DHA.

8.4.1 Department of Home Affairs

The control, regulation and facilitation of immigration and movement of persons through the ports of entry and determination of the status of asylum seekers and refugees in accordance with international obligations is the mandate of the Department of Home Affairs. In its duty the Department assesses if the persons wishing to enter the country are desirable or undesirable in terms of the Immigration Act. In performing this function the Department plays a crucial role in national security of the country as it able to exclude or prohibit those that may fall under any of these categories: infected with infectious diseases that can spread easily; fugitive from justice – those that have warrant of arrest against the person or a conviction for genocide, torture, drug trafficking, money laundering, kidnapping, terrorism, or murder; members or supporters of an organisation practising racial hatred or social violence; members of an organisation using crime or terrorism to reach its goals; people who have previously been deported and have not been rehabilitated by the Department in the prescribed manner.

The Department of Home Affairs seeks to ensure that immigration is managed effectively and securely in the national interest including economic, social and cultural development. The immigration policy as prescribed by the Immigration Amendment Act, 2004 (Act 19 of 2004), discourages illegal migration into South Africa by encouraging foreign nationals to apply for different permits to legalise their stay in the country while at the same time seeks to create an enabling environment for foreign direct investment in South Africa and attract scarce skills required by the economy in accordance with the 2014 vision of eradicating poverty and underdevelopment.

8.4.2 Local Government (City of Johannesburg)

All the migrants that have been processed by the Department of Home Affairs or those that are undocumented live in local communities. The city plays an unavoidable frontline role in managing the integration of all classes of migrants. Migrants – including irregulars – consume services, participate in the informal economy and are residents of socially excluded areas targeted for assistance. Cross border migrants are accused by certain organised groupings – rightly or wrongly –of being illegitimate competitors for scarce social resources, including low-/semi-skilled employment and opportunity to trade. The conflicts arising from these accusations, and their consequences, occur and must be managed at the local level, as they starkly did (and were) in the wave of recurring xenophobic attacks directed against internal and cross-border migrants living in Johannesburg's disadvantaged communities that been arguably occurring since 2008.

The city's approach to date has been a combination of direct advice to migrants with far-reaching dialogue and lobbying. However, the challenge of facilitating access to services for migrants remains and so are the broader issues pertaining to tensions between migrants and locals, which, if not managed, would place the city at the centre of a potential social and humanitarian crisis. A much more engaged and aggressive migrant integration policy is required to manage this complex set of issues proactively.

The Refugee Act states that all refugees are entitled to health care, to seek employment and to education in the same way as South African citizens. It also states that all people in the country are entitled to the rights enshrined in Chapter 2 of the Constitution, with the exception of political rights and the rights to freedom of trade, occupation and profession, which do not apply to non-citizens. Legal immigrants and refugees are, therefore, entitled to services offered at the municipal level such as safety, housing, clinic services, libraries, etc. Indeed, in some instances, non-nationals are actively denied these services. The city policy thus seeks to put plans in place so that services are extended to refugees, asylum seekers and other legal immigrants. The systems that are to be set up in terms of the policy would thus ensure that access for refugees and other migrants is actively facilitated, such as providing translation services or culturally sensitive versions of municipal services.

On the question of irregular migration, the city has neither an unambiguous enforcement nor a clear and formalised regularisation role to play. Many of the more complex counter-xenophobia challenges relate to the increasing participation of irregular migrants in the social and economic space of the city, and the city clearly has a role (from a public safety point of view) in preventing any violent or otherwise intimidating illegal activities conducted against them, particularly where such activities take on an organised form. But the city is not, and will never be, an immigration authority and cannot of its own accord provide pathways to regularisation.

8.4.3 The United Nations High Commissioner for Refugees (UNHCR)

The UNHCR helps the Government of South Africa clear the backlog of pending asylum claims and ensure a more rapid review of new claims. It also contributes to creating an environment conducive to the local integration of refugees and provides limited humanitarian assistance to vulnerable asylum seekers and refugees, especially women and children. It helps the government, NGOs and its partners to build capacity to respond more effectively to refugee needs, protect and assist refugees and asylum seekers in accordance with international standards. This includes the provision of support and training to the Department of Home Affairs (DHA). It also ensures that refugees and asylum seekers have access to national social services, including education, health and assistance programmes and mobilises resources jointly with partners for refugee assistance, while also facilitating durable solutions for refugees, including local integration for those with limited prospects of returning home, using resettlement as a protection tool for those with protection needs, and aiding voluntary repatriation in safety and dignity.

8.4.4 International Organization for Migration (IOM)

The International Organization for Migration (IOM) is the leading intergovernmental organisation in the field of migration and is committed to the principle that humane and orderly migration benefits migrants and society. The IOM is part of the United Nations system, as a related organisation. The IOM supports migrants, developing effective responses to the shifting dynamics of migration and, as such, is a key source of advice on migration policy and practice. The organisation works in emergency situations, developing the resilience of all people on the move, and particularly those in situations of vulnerability, as well as building capacity within governments to manage all forms and impacts of mobility. The organisation is guided by the principles enshrined in the Charter of the United Nations, including upholding human rights for all. Respect for the rights, dignity and well-being of migrants remains paramount. The IOM in South Africa works to coordinate responses at national, regional and international level with respect to supporting reconstruction efforts, demobilisation of former combatants, and reintegration of Internally Displaced Persons (IDPs) and the repatriation of refugees.

8.5 Summary

The magnetic effect of eGoli in attracting people within and beyond its borders persists even in contemporary times. The dawn of democracy somehow lifted the veil which limited migration to contract labourers from the region and skilled manpower from abroad. Johannesburg has maintained its trademark among immigrants as the first choice of destination within the region and for those coming from abroad. The disbandment of draconian apartheid immigration policies and legislations opened the country to immigrants who see the city and the country at large to be full of opportunities. This, coupled with favourable migration policies, has led to an influx of both legal and illegal migrants which the country is struggling to control. Its porous borders and corruption at points of entry have worsened the situation as most illegal migrants are from within the region – a factor which is cause societal attrition due to competition for opportunities.

Notes

1. Section 26 of the Constitution provides that 'everyone has the right to access to adequate housing'. Section 26(2) confers a duty upon the State to progressively facilitate access to adequate housing, within its available resources. The Housing Act No. 107 of 1997, as amended ('Housing Act') was enacted to give effect to the Constitutional right to access to housing. In terms of section 1 of the Housing Act, this right will apply on a progressive basis to 'all citizens and permanent residents of the Republic'. Notwithstanding the above, the Housing Act therefore does not grant refugees or temporary residents the right to access housing.
2. Constitution sect 9(1) & (2).
3. Constitution, sects 25, 26 & 27.
4. Refugee Act 130 of 1988.
5. National Consortium for Refugee Affairs 'Summary of key findings: Refugee protection in South Africa (2006) *Kutlwanong Democracy Centre*.
6. National Consortium for Refugee Affairs.
7. Refugee Act, art 27(b).
8. Refugee Act, art 27(f) & (g).
9. Refugee Act, art 34.
10. Refugee Act, art 2.
11. National Consortium for Refugee Affairs.
12. National Consortium for Refugee Affairs.
13. Universal Declaration of Human Rights (UDHR) 1948 art 13(2), International Convention on the Protection of the Rights of All Migrant Workers and Members of Their Families adopted on 18 December 1990; UN Convention and Protocol Relating to the Status of Refugees adopted by General Assembly Resolution 2198 (XXI) 1951 art 1(2).
14. Constitution sects 27(1)(c) & 28(1)(c).

Bibliography

Aboobaker, S. 2015. Foreign-Owned Shops in Alexandra Looted. Available From https://www.iol.co.za/news/south-africa/gauteng/foreign-owned-shops-in-alexandra-looted-1847071#.Vt36fsgqqko

Abrahams, C. & Everatt, D. 2019. City Profile: Johannesburg, South Africa. *Environment And Urbanization Asia*, 10, 255–270.

Abrahams, C., Everatt, D., Van Den Heever, A., Mushongera, D., Nwosu, C., Pilay, P., Scheba, A. & Turok, I. 2018. South Africa: National Urban Policies and City Profiles for Johannesburg and Cape Town. Johannesburg. Available From http://www.centreforsustainablecities.ac.uk

Amit, R. 2010. *Lost in the Vortex: Irregularities in the Detention and Deportation of Non-Nationals in South Africa*. Report for FMSP. Johannesburg: Wits University.

Chipkin, C. M. 1993. *Johannesburg Style: Architecture & Society, 1880s–1960s*. David Philip Publishers.

Cohen, R. 2008. *Global Diasporas: An Introduction*. Routledge.

CoJ. 2011. *Integrated Development Plan (IDP)*. Johannesburg: City Of Johannesburg Press.

Craissati, D., Banerjee, U. D., King, L., Lansdown, G. & Smith, A. 2007. *A Human Rights Based Approach to Education for All.* UNICEF.

Crush, J., Chikanda, A. & Skinner, C. 2015. *Mean Streets: Migration, Xenophobia and Informality in South Africa.* African Books Collective.

Crush, J. & Tawodzera, G. 2011a. Medical Xenophobia: Zimbabwean Access to Health Services in South Africa.

Crush, J. & Tawodzera, G. 2011b. Right to the Classroom: Educational Barriers for Zimbabweans in South Africa. In *Southern African Migration Programme: Open Society Initiative for South Africa.* Migration Policy Series No 56. Cape Town: Institute for Democratic Alternatives in South Africa.

Crush, J., Williams, V. & Peberdy, S. 2005. Migration in Southern Africa. In *Policy Analysis and Research Programme of the Global Commission on International Migration.* Global Commission on International Migration.

Datta, K., McIlwaine, C., Evans, Y., Herbert, J., May, J. & Wills, J. 2006. *Work and Survival Strategies among Low-Paid Migrants in London.* London: Queen Mary, University of London.

De Klerk, N. 2015. Looters Target Foreign-Owned Shops in Joburg. *News24.*

Dodson, B. 2010. Locating Xenophobia: Debate, Discourse, and Everyday Experience in Cape Town, South Africa. *Africa Today,* 56, 2–22.

The Economist. 2013. Hot spots 2025. Benchmarking the future competitiveness of cities. The Economist Intelligence Unit Limited 2013.

Euromonitor. 2017. *City Review: Johannesburg City Review.* Euromonitor International.

Fatti, C. E. & Vogel, C. 2011. Is Science Enough? Examining Ways of Understanding, Coping with and Adapting to Storm Risks in Johannesburg. *Water SA,* 37, 57–65.

Greenburg, J. & Polzer, T. 2008. Migrant Access to Housing in South African Cities. In *Migrants' Rights Monitoring Project.* Johannesburg: Witwatersrand University. (Forced Migration Studies Programme Special Report No 2).

Gushulak, B. D. & Macpherson, D. W. 2006. *Migration Medicine and Health: Principles and Practice.* Pmph-Usa.

Hacker, K., Anies, M., Folb, B. L. & Zallman, L. 2015. Barriers to Health Care for Undocumented Immigrants: A Literature Review. *Risk Management and Healthcare Policy,* 8, 175.

Hawker, D. 2015. South African Businesses Hurt by Jeppestown Looting. https://www.enca.com/south-africa/south-african-businesses-affected-jeppestown-looting

Hirschman, C. 1982. Immigrants and Minorities: Old Questions for Mew Directions in Research. *International Migration Review,* 16, 474–490.

Hlubi, P. 2015. Foreign-Owned Shops in Snake Park and Doornkorp were Targeted by Looters. *Enews,* 20 January 2015.

Holland, H. & Roberts, A. 2002. *From Jo'burg to Jozi: Stories about Africa's Infamous City.* Penguin Global.

Kabwe-Segatti, A. W. 2008. *Migration in Post-Apartheid South Africa: Challenges and Questions to Policy-Makers.* Agence française de développement (Afd), Département de la Recherche.

Knox, P. & Gutsche, T. 1947. *Do You Know Johannesburg? [With Illustrations And A Street Plan.].* Unie-Volkspers.

Laher, H. 2008. *Antagonism toward African Immigrants in Johannesburg, South Africa: An Integrated Threat Theory (ITT) Approach.* University of the Witwatersrand Johannesburg.

Landau, L. & Polzer, T. 2007. *Xenophobic Violence, Business Formation, and Sustainable Livelihoods: Case Studies of Olievenhoutbosch and Motherwell*. Johannesburg: University of the Witwatersrand. Forced Migration Studies, University of Witwatersrand, 1–23.

Landau, L. & Polzer, T. 2008. Working Migrants and South African Workers: Do they Benefit Each Other? *South African Labour Bulletin*, 32, 43–45.

Landau, L. B. & Jacobsen, K. 2004. Refugees in the New Johannesburg. *Forced Migration Review*, 19, 44–46.

Macpherson, D. W., Gushulak, B. D. & Macdonald, L. 2007. Health and Foreign Policy: Influences of Migration and Population Mobility. *Bulletin of the World Health Organization*, 85, 200–206.

Maki, H. 2010. Comparing Developments in Water Supply, Sanitation and Environmental Health in Four South African Cities, 1840–1920. *Historia*, 55, 90–109.

Mandy, N. 1984. *A City Divided: Johannesburg and Soweto*. New York: St. Martin's Press.

Martinez, R., Buntin, J. T. & Escalante, W. 2012. The Policy Dimensions of the Context of Reception for Immigrants (and Latinos) In The Midwest. Cambio De Colores (10th: 2012: Kansas City, Mo.). Cambio De Colores: Latinos in the Heartland: Migration and Shifting Human Landscapes: *Proceedings of the 10th Annual Conference: Kansas City*, Missouri, June 8–10, 2011. Columbia, MO: University of Missouri. Cambio Center.

Mbhele, M. S. 2016. *Exploring Schooling Experiences and Challenges of Immigrant Learners in a Multilingual Primary School*. Durban: University of KwaZulu Natal.

McConnell, C. 2009. Migration and Xenophobia in South Africa. *Conflict Trends*, 2009, 34–40.

Monson, T., Landau, L., Misago, J. & Polzer, T. 2010. May 2008 Violence Against Foreign Nationals in South Africa: Understanding Causes and Evaluating Responses.

Mthembu-Salter, G., Amit, R., Gould, C. & Landau, L. B. 2014. *Counting the Cost of Securitising South Africa's Immigration Regime*. Sussex: Sussex University.

Murray, M. J. 2011. *City of Extremes: The Spatial Politics of Johannesburg*, Duke University Press.

Northcote, M. A. 2015. *Enterprising Outsiders: Livelihood Strategies of Cape Town's Forced Migrants*. Ontario: The University of Western Ontario.

Ogbu, J. U. & Matute-Bianchi, M. E. 1986. Beyond Language: Social and Cultural Factors in Schooling Language Minority Students. In J. Perlmann and H. Vermeulen (Eds) *Immigrants, Schooling, and Social Mobility: Does Culture make a Difference*. Los Angeles, CA: Evaluation, Dissemination, and Assessment Center, Bilingual Education Office.

Okyere, D. 2018. Economic and Social Survival Strategies of Migrants in Southern Africa: A Case Study of Ghanaian Migrants in Johannesburg, South Africa.

Oosthuizen, M. & Naidoo, P. 2004. Internal Migration to the Gauteng Province. Development Policy Research Unit. Working Paper 04/88, University of Cape Town.

Oxford Economics. 2016. Global Cities 2030. Future trends and market opportunities in the world's largest 750 cities. How the global urban landscape will look in 2030. Oxford Economics.

Parilla, J. & Trujillo, J. 2015. South Africa's Global Gateway: Profiling the Gauteng City-Region's Interntional Competitiveness and Connections. In *Global Cities Initiative, A Joint Project of Brookings and JP Morgan Chase*. Global Cities Initiative.

Peberdy, S. 2013. Gauteng: A Province of Migrants. *Gauteng City-Region Observatory Data Brief.*
Peberdy, S. 2016. *International Migrants in Johannesburg's Informal Economy.* African Books Collective.
Polzer, T. 2010. *Migration Fact Sheet 1: Population Movements in and to South Africa.* Johannesburg: FMSP.
Robertson, C. C. 1986. *Remembering Old Johannesburg.* Ad. Donker.
SACN. 2018 *State of City Finances Report 2018.* Johannesburg: South African Cities Network (SACN).
Schäffler, A., Christopher, N., Bobbins, K., Otto, E., Nhlozi, M., De Wit, M., Van Zyl, H., Crookes, D., Gotz, G. & Trangoš, G. 2013. State of Green Infrastructure in the Gauteng City-Region. In *Gauteng City-Region Observatory (Gcro), A Partnership of The University of Johannesburg.* The University of the Witwatersrand, Johannesburg, and the Gauteng Provincial Government.
Shorten, R. 1966. *The Johannesburg Saga.* Johannesburg: John R. Shorten Propriety Limited.
Smit, R. & Rugunanan, P. 2014. From Precarious Lives to Precarious Work: The Dilemma Facing Refugees in Gauteng, South Africa. *South African Review of Sociology,* 45, 4–26.
Stats SA. 2015. Census 2011: Migration Dynamics In South Africa. In *Statistics South Africa.* Pretoria: Statistics South Africa.
Storie, M. 2014. Changes in the Natural Landscape. Johannesburg: Changing Space Changing City: Johannesburg after Apartheid. Wits University Press, 137–153.
Suphanchaimat, R., Kantamaturapoj, K., Putthasri, W. & Prakongsai, P. 2015. Challenges in the Provision of Healthcare Services for Migrants: A Systematic Review through Providers' Lens. *Bmc Health Services Research,* 15, 1–14.
Symonds, F. A. 1953. *The Johannesburg Story,* F. Muller.
Teixeira, C. & Halliday, B. 2010. Introduction: Immigration, Housing and Homelessness. *Canadian Issues,* 3.
Tomlinson, R., Beauregard, R., Bremmer, L. & Mangcu, X. 2003. *Emerging Johannesburg.* New York and London: Routledge.
Turok, I. 2012. *Urbanisation and Development in South Africa: Economic Imperatives, Spatial Distortions and Strategic Responses.* Human Settlements Group, International Institute for Environment And.
United Nations. 2017. International Migration 2017. Department of Economic and Social Affairs: Population Division. Accessible online: https://www.un.org/en/development/desa/population/migration/publications/wallchart/docs/MigrationWallChart2017.pdf
Vandeyar, S. 2010. Educational and Socio-Cultural Experiences of Immigrant Students in South African Schools. *Education Inquiry,* 1, 347–365.
Vearey, J. & Nunez, L. 2010. Migration and Health in South Africa. *Background Paper and Report on the National Consultation on Migration and Health in South Africa, Midrand, 22nd–23rd April.*
Walls, H. L., Vearey, J., Smith, R. D., Hanefeld, J., Modisenyane, M., Chetty-Makkan, C. M. & Charalambous, S. 2016. Understanding Healthcare and Population Mobility in Southern Africa: The Case of South Africa. *South African Medical Journal,* 106, 14–15.
WHO. 2010. Health of Migrants: The Way Forward: Report of a Global Consultation, Madrid, Spain, 3–5 March 2010.

9 Lusaka

A Retreat into Zambia

9.1 Introduction

Lusaka is the capital city of Zambia, a country in the Central African Plateau with an average altitude of 1,000–1,400 m above sea level. Zambia is generally considered to be a Southern African country, because of its strong social and economic ties with the countries in the Southern African sub-continent rather than those in Central and Eastern Africa. Zambia lies between latitudes 100° and 180° south and 220° and 330° east. It is landlocked and shares borders with eight neighbouring countries (Figure 9.1). Zambia has a land area of 752,614 km² and a population of just over 10 million (Mulenga, 2003).

The development of Zambia's urban centres can only be traced back as far as the annexation of the territory now called Zambia by the British South

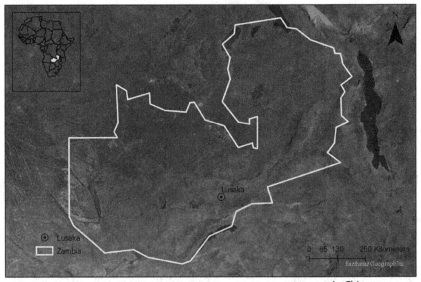

Figure 9.1 Map of Zambia Showing the Position of Lusaka
DOI: 10.4324/9781003184508-9

African Company (BSAC) between 1899 and 1900. Before that most of the territory was under the control of different African Kingdoms. The people throughout the territory, however, lived in fear of the so-called slave traders, who attacked communities, enslaved their captives and sent them to the east or west coasts through present-day Tanzania, Mozambique and Angola, respectively, for onward shipment to the Middle East, Western Europe and South America as slaves. Thus, at the time when Zambia was brought under European colonial rule, there were no major urban centres, perhaps because the activities of the slave raiders had destroyed stable communities and their economies. Zambian cities and towns thus date back only to the early 1930s, while the country's urbanisation has largely resulted from the copper mining industry, which transformed the economy from a stagnant one based on labour migration to the South African and Southern Rhodesian mines and farms in the early years of colonial rule to a vibrant economy based on a growing mining industry in the post Second World War period. However, Zambia's mining industry, which has been the backbone of the national economy, began to stagnate and decline in the mid-1970s. The situation worsened in the 1980s due to low copper prices, declining production and a rising debt burden. The pace of urbanisation in Zambia has, however, generally mirrored the economic trends in the dominant copper mining industry (Hall, 1965; Gann, 1964).

Most of the cities and towns in Zambia emerged in two zones – firstly, along the railway line that was constructed for the purpose of connecting the rich copper mines in the Katanga region of the then Belgian Congo to the South African ports, and secondly, on the Copperbelt, where towns and cities emerged around the copper mines. Other towns also emerged around administrative centres that were established for administering the large sparsely populated territory. The urban population in Zambia is, however, concentrated in the cities and towns of the Copperbelt. The towns around the copper mines grew most rapidly between 1945 and the immediate post-independence period. Lusaka, however, has been the fastest growing city in the post-independence period. This can be attributed to its role as the administrative centre of a newly independent country whose leaders were keen to play a role in international affairs. As a result, Lusaka had to provide facilities to hold international conferences, such as the summit for the head states of the Non-Aligned Movement held in Lusaka in 1970, followed by the Commonwealth Heads of State summit in 1979.

Lusaka started as a railway siding in 1905, when the railway line that was constructed primarily to transport copper from Katanga Province in the present-day Democratic Republic of Congo to the seaports of South Africa reached Lusaka. Within a few years, however, Lusaka attracted a number of white settler farmers, mostly of Afrikaner origin. The BSAC, which was administering the territory at the time on the basis of a royal charter obtained from the British monarchy, was compelled to grant the growing white population around Lusaka with the right to manage their local affairs. An elected Village Management Board was thus established in 1913 with the mandate of managing the affairs of an emerging service centre. The original size of Lusaka, which was under the jurisdiction of the Village Management Board was a narrow strip of land along the railway line. It was 5 km in length and 1.5 km wide, with the railway being the centre of the area under the jurisdiction of the Lusaka Village Management Board (Williams, 1984). The city has since been extended to 360 km^2, while a recently completed integrated development plan proposes to extend the city boundary beyond this to bring the Lusaka International Airport and a substantial amount of rural land within the city boundary. The proposed extension if approved will bring a substantial amount of land within the city boundary and probably help resolve the current shortages of land for burial sites and disposal of the city's solid waste.

The rapid growth of Lusaka, however, began in earnest in 1931, when it was designated as the new capital or principal administrative centre of Northern Rhodesia, as Zambia was then called. Its selection as the new capital was due to its central location on the main north-south axis of the railway line, which was expected to become the centre of development. The central location of Lusaka was also evident from being the intersection of the main roads to the north and south, and east and west. Lusaka was also within easy reach of the Copperbelt, the country's economic heartland. Furthermore, unlike other equally central locations, Lusaka had substantial underground water resources in its limestone/dolomite aquifers, which could provide the city with adequate water throughout the year. Lusaka is thus a planned city. Its original plan was made by Professor Adshead, who conceived the city of Lusaka as an administrative centre only. The original plan did not therefore provide for economic activities other than government administration, domestic and menial services. Industrial activities and a large population of Africans were in particular not anticipated to form a part of the city of Lusaka (Collins, 1969).

Industrial activities, however, became part of the Lusaka Development plan during its implementation. The government planner responsible for implementation of the plan, J. T. Bowling, modified the original plan and provided for light and heavy industrial areas. The area between Church Road, the railway line and the Great East Roads was, for example, zoned for light industrial activities, while heavy industrial activities were allocated the immediate western side of the shopping area, which had emerged to the west of the railway line. The shopping area, on the other hand, was designated to move to the hilly and more attractive, well-drained land on the Ridgeway, southeast of the railway line.

The principal planners of Lusaka never intended it to be a large city. Its initial total area was only 2.6 km^2. It was, however, increased to 18 km^2 in 1931, then 36, and 139 and 360 km^2 in 1961, 1969 and 1970, respectively (Collins, 1969; Pasteur, 1982). The Integrated Development Plan for the city completed in 2000 also proposed extension of the city boundary to bring the Lusaka International Airport, which is currently in the neighbouring Chongwe District and additional land around the city within its boundaries. Inadequate land within the city boundary has constrained the re-development of the slum areas, which initially emerged as unauthorised areas, but most have since been recognised as improvement areas under the Improvement Areas Act of 1974. There is also a shortage of land for new burial and solid waste disposal sites within the current city boundary as the existing burial and solid waste disposal sites have become full. The land shortages being experienced within the existing city boundary are partly due to the dramatic population increase in the city.

The economy of Lusaka has become more diversified with its physical expansion and population growth. It has in fact grown from the provision of a few services to commercial farmers who had established themselves around it to the provision of higher order services, such as financial and technical services, construction and even manufacturing activities. As the capital city of Zambia, Lusaka also provides services, including administrative functions to Zambia as a whole. However, Lusaka also plays a significant role in the country's manufacturing. Most manufacturing enterprises are located in Lusaka and the Copperbelt. Food processing enterprises, such as milling, meat processing and production of essential commodities such as detergents and other domestic chemical products seem to be concentrated in Lusaka. In terms of employment, the service sector is the largest employer of the city's labour force. Services and administration in particular have consistently accounted for most of the formal employment in the city. This suggests that though the economy of the city is more diversified than that of the country, it is quite weak, as most of the sectors are underdeveloped. The basic manufacturing activities, such as food processing and beverages, textiles and leather goods, for example, dominate the manufacturing activities. The transport and communication and hotel and restaurant sectors are also underdeveloped and perhaps have scope for improvement. The underdevelopment of these sectors could be attributed to the internal orientation of the Zambian economy prior to the recent macroeconomic reforms.

The performance of the construction sector, on the other hand, has mirrored the performance of the overall economy. The construction sector has, therefore, performed strongly during periods when the national economy has been buoyant and declined during the years of economic stagnation and decline. Although the statistics are not available, the construction sector in Lusaka performed relatively well in the late 1990s. It was in particular helped by the rehabilitation of the major roads and the construction of new housing estates and conference facilities (Seshamani, 2002). The primary economic activities, especially agricultural and mining activities have been on the decline, as more of the city's land has become built up. The financial and commercial sectors, on the other hand, are fairly large and account for most of the financial and commercial activities in the country.

Although the economy of the City of Lusaka is somewhat more diversified than the national economy, like the national economy, it only provides formal employment to a small proportion of its labour force. The Integrated Development Plan for Lusaka, for example, put the number of people in formal employment in Lusaka at 120,233 or 35% of the labour force (V3 Consulting

Engineers, 2000). The majority (65%) of the city's labour force, therefore, earns its livelihood from informal economic activities, which predominantly consist of unregistered and unregulated small-scale non-agricultural economic activities ranging from petty trading to metal fabrication and wood processing. The bulk of the informal economic activities are, however, essentially in trading. The low proportion of the labour force working in the formal sector has a bearing on welfare of the residents of the city and will be discussed later in the chapter, which analyses slums and welfare.

9.2 Migrants Profile in Lusaka, Zambia

According to Monu et al. (2012), Lusaka City is home to people from different ethnic groups, nationalities and religions. Being the capital city of Zambia, people from different parts of the country were attracted to the City because of the opportunities that were available. According to 'Lusaka City' (2013), Lusaka City was primarily an area of the Soli ethnic group. However, different people from different parts of the country with different cultures entered the City and diluted the Soli culture, especially during the post-1964 (independence) years. Lusaka City ceased to be a Soli ethnic group area and became a home for the different ethnic groups. In terms of language, English is the official language of Zambia and was spoken in Lusaka City. As for the indigenous languages, Mwakikagile (2010) pointed out that the Nyanja language was predominantly spoken due to the immigration of people from the Eastern Province of Zambia, where the Nyanja language was spoken. Second to Nyanja was the Bemba language which was widely spoken in the City. This was also due to the movement of people from the Northern, Luapula, Copperbelt and Central Provinces where the Bemba language, or at least its dialect, was spoken.

Zambia is quite open to foreigners: on average, just 5.8% of applicants were denied admission over the 2009–2012 period. China, India and South Africa supply the largest numbers of migrants to Zambia, followed by Zimbabwe, the United Kingdom and the United States. Migrants from the United States and United Kingdom enter on short-term work permits and leave after the duration of their contract. Indians, South Africans and Zimbabweans tend to migrate more permanently. Consensus on the Chinese was divided. Migration is a significant contributor to urban growth and the urbanisation process, as people move in search of social and economic opportunities and as a result of environmental deterioration (Awumbila, 2017).

Immigrants to Zambia in 2019 numbered 170,200. The majority of these individuals were from the Democratic Republic of Congo, followed by those from Angola. The 2010 national census of population and housing in Zambia is one of the primary sources of migration data; it counted an immigrant stock of 43,867, representing 0.4% of the total population. Of that, the gender split of immigrants was 53% male and 47% female. The top 10 sources of immigrants are the Democratic Republic of the Congo, Angola, Zimbabwe, India, China, Rwanda, the United Republic of Tanzania, South Africa,

Somalia, the United Kingdom and Burundi. According to UN DESA, in 2019 immigrants from the Republic of Congo and from Malawi also formed a significant proportion of Zambia's international migrant stock.

The total number of registered asylum seekers was 2,533 in the period of 2013–2017, 61% males and 39% females. During the same period, 15,933 refugees (54% males and 46% females) were registered in Zambia. Of the registered asylum seekers and refugees, the majority were from the Democratic Republic of the Congo, while some of them originated from Burundi and Somalia. There was a new flow of refugees into the country towards the end of the second quarter of 2017. Between August 2017 and June 2018 there were a total of 13,753 new arrivals (refugees), and most of them were from the Democratic Republic of the Congo. In 2018 alone 9,233 new refugee arrivals were recorded, and at the end of 2019 there were 7,870 new arrivals, all with access to refugee status determination procedures.

The well-being of migrants largely depends on the availability of work generating a decent income, a clear and secure legal status, access to social services, access to social and health protection, and their participation in society. Together with a decent job and a decent income, a clear and secure legal status is a critical component of the social condition and well-being of migrants. A temporary residence permit or work contract is normally sufficient to provide legal security to short-term migrants (UN DESA, 2004).

In Zambia, 70% of the urban population resides in slums characterised by poor dilapidated housing and other basic support infrastructure such as reticulated water and sewer. The urban housing deficit is estimated to be 1,539,000 units, that is expected to reach over 3.3 million by 2030 nationally (Centre for Affordable Housing in Africa, 2017). Slum dwellers in Zambia to a greater extent represent the poorest echelons of the population, and since they live below the poverty datum line, they cannot afford to incrementally improve their housing conditions due to the relatively high construction costs. Even then, they are systematically marginalised from conventional micro finance because of their meagre financial resources, as well informal sources of income and informal tenure security which they cannot use as collateral from formal institutions. Migrants interviewed in this study expressed that migrant population face a moderate degree of challenges in accessing housing. Their situation is almost the same as the local citizens.

Employment is one of the main motives for migrants to come to Zambia, as evidenced by the fact that the Employment Permit constitutes the highest number of permits issued by the Department of Immigration. More than 65% of permits issued by the Department of Immigration were work-related permits (Employment and Temporary Employment permits). Employment Permits were the most commonly issued at 49.4% of total permits issued, followed by Temporary Employment Permits at 19.5% (IOM, 2019). This is the only legal basis on which migrants can participate in economic activities as granted in the permit; otherwise, it becomes illegal to engage in economic activities other than those granted in the permit. Interviews conducted with migrants in this study reveal that migrants have less challenges in accessing economic challenges in Lusaka. The copper industry plays a central role in the Zambian economy, accounting for a large part of the total exports and GDP. The mining sector has attracted labour migration directly by creating job opportunities and indirectly by contributing to the economic development of the mining cities and the economy as a whole (Coderre-Proulx et al., 2016). All things being equal, if these migrants are engaged in formal economic activities, they will contribute to the economy in taxes and social involvement. Also, immigrants in formal employment have equal access to contributory social protection, including the National Pension Scheme Authority and Workers' Compensation Fund. Other forms of social protection such as access to the Social Cash Transfer and the Public Welfare Assistance Scheme programmes are only available to nationals. However, refugees residing in Lusaka due to various reasons mostly live in compounds. Their income is very limited due to restrictions in relation to access to wage/ formal employment and self-employment. They struggle to be self-reliant. If their migration is well managed, they can also contribute to skills transfer within Zambia.

Chinese migration to Zambia has increased in recent years following the development of a strong economic relationship between the two countries, and against a backdrop of rising Chinese migration to resource-rich areas of the world. China has invested billions of dollars in Zambia's most profitable industries and is responsible for most of the major infrastructure projects in the country. Flows of people have begun to follow the flows of investment capital: the number of Chinese nationals entering Zambia has increased by over 60% since 2009. The Chinese flow to Zambia, characterised by highly skilled migrants, exemplifies the phenomenon of South–South migration, which represents more than one-third of global migration.

Zambia has a non-discriminatory approach to accessing health care services for all people in the country. The Constitution of Zambia and the National Health Policy promote an equitable provision of health care services across the population. In addition, the Zambia National Health Act provides for the right to health care for all.

Education and migration are linked by a complex two-way relationship. The first kind of interaction consists of the various direct and indirect impacts that education has on migration. Education is recognised as a driver of migration as it creates openness to as well as opportunities for employment abroad. The second kind of interaction consists of the many ways in which migration impacts education in both the origin and the destination populations. Not all migrants settle for a lifetime in the destination country; some migrants who leave for educational reasons return to their homeland and others remain in the country to which they migrated. Return migrants bring back to their home country an experience and education that they gained in the host country, a mechanism by which international migration contributes to building human capital (Fargues, 2017).

The right to education is not a constitutionally justiciable right, but there is provision for free primary education. Bursaries are available for children unable to access secondary and tertiary education. As a signatory to the 1951 Refugee Convention, Zambia has a reservation to Article 22 of the convention on education. Education facilities in the settlement/refugee camp are provided by the government and the UNHCR. Unfortunately, only basic education is available. The UNHCR does not have funds for secondary education and most refugees (especially those in the settlements) do not benefit from secondary education. The UNHCR usually commits itself to offering educational support (to cover things like uniforms, books and so on) to refugees living in local communities with valid urban residence. For immigrants living outside the settlements and without legal authorisation, there is no educational support, which means that many of these children do not attend school. Most of them struggle to enrol in government schools, which de facto limits their access to free, state primary education. As a consequence of this, refugees and other migrants in urban areas usually take their children to community or private schools whose fees are barely affordable (Chigavazira et al., 2012).

Generally, Zambia has a history of peaceful coexistence with other nationals in the communities. However, periodically, sporadic acts of xenophobic violence have flared up (Nthanda, 2016). The Department of Immigration, in conjunction with other security wings of the Government of Zambia, such as the Zambia Police Service, conducts periodic operations to identify irregular immigrants within workplaces and residential areas of Zambia (IOM, 2019).

Social integration is a process by which migrants are integrated into host communities and are issued with documentation, such as Residence Permits, which allows them to enjoy many of the same rights as citizens. Furthermore, Zambian law makes provision for non-nationals to acquire citizenship through various means such as having been resident in Zambia over a continuous period of ten years (Chigavazira et al., 2012). While Zambia has generally remained open and hospitable to refugees, until recently it required them to remain in camps or settlements. Up until 2017, the reception of refugees was governed by the 1971 Control Act; as the name suggests, the concern was control of refugees. Though Zambia is a signatory to the 1951 Refugee Convention, it did not adopt most of its refugee-rights provisions into national laws. Recently, the Zambian government has grown more willing to grant freedom of movement to refugees, given residency rights to former refugees from Angola and Rwanda, and committed to a new form of settlement for newly arrived refugees from the DRC. This restrictive national refugee policy was applied to varying degrees across the state. It is a stark contrast to the global trend of reducing refugee protection, evident in Europe, Australia and even Zambia's neighbour, South Africa. And it's a change for Zambia itself. In Zambia, as in neighbouring Tanzania and Uganda, many refugees have found de facto integration in border towns and urban areas, with government officials turning a blind eye to their presence. Yet such a laissez-faire approach often leaves those refugees without official permission to live outside a camp in fear of detention, deportation or extortion by corrupt immigration officers.

Lately, however, something appears to have shifted. Firstly, Zambia committed to integrating former Angolan and Rwandan refugees. In 2014, the state created the Strategic Framework for the Local Integration of Former Refugees in Zambia (SFLI), which aimed to regularise the status of 10,000 former Angolan refugees and 4,000 former Rwandan refugees. While there are still issues with implementation, this approach to former Angolan refugees sits in stark contrast to their Southern Africa neighbours, South Africa. There has also been significant international pressure on Rwandan refugees to repatriate, but Zambia has resisted this and tried to grant forms of permanent residence to this population from Rwanda.

Though migrants from neighbouring African countries (apart from those mentioned above) do not figure prominently in permit and visa data, this is likely because they do not enter Zambia through formal channels. Such data are collected manually and inconsistently at border posts, and there is no enforcement along the majority of the border. The Zambian government is largely unaware of the volume or characteristics of migrants crossing into the country informally, according to the Deputy Chief of Operations of the Department of Immigration. Though Somali and Congolese immigrants have built a sizeable community in Lusaka, Zambia is still mainly a transit point for Central and East African migrants traveling to South Africa; many do not choose to reside in Zambia permanently.

In some countries, 'migrants ... are virtually indistinguishable from the receiving population', according to Castles. This is true in Zambia, where South African, Zimbabwean and British immigrants share many cultural, occupational and physical similarities with native Zambians. Due to the strong colonial and regional ties among these countries, it can be impossible to discern an individual's nationality. India and Zambia also have a long migration and shared cultural history as former British colonies; many Indian migrants have so assimilated into Zambian society that they are identified as Zambians.

Chinese migrants are more conspicuous and less assimilated into Zambian society than other migrant groups, at least partially due to the sheer newness of their presence. They also attract disproportionate attention due to their work on high-profile projects in visible industries such as construction, and a distinct appearance relative to most Zambian residents. Though no Zambian integration initiatives exist for any migrant population, the Chinese have remained by far the most segregated. Many attribute this to persistent language and cultural differences. Though an increasing number of Chinese migrants speak English, few Zambians are proficient in Chinese. However, the Chinese-sponsored introduction of Mandarin instruction in Zambian government secondary schools in 2014 has the potential to close this language gap.

Some Zambians praise the Chinese for living and working side-by-side with locals, rarely seen amongst other groups of expatriates. Others, however, are angered by what they view as Chinese migrants' perceived sense of superiority, exploitation of local workers and unwillingness to learn local languages. Cultural differences drive misunderstandings on both sides: the Chinese consistently complain about Zambians' work performance, calling them lazy, while Zambians are mystified by the Chinese tendency to migrate without their families, feeling this signals a cold and unapproachable temperament.

The Chinese straddle a visible divide in the foreign populations in Zambia. More permanent migrants, mainly from other African countries such as South Africa, Zimbabwe and Somalia, but also Indians and Lebanese, tend to interact on a peer-to-peer level with local Zambians. They are more established and integrated in-country. Expatriates are the second major group, mainly Americans and Europeans on contract with international organisations, governments or religious organisations. It is highly unusual for anyone from this demographic to remain in Zambia longer than two years. The fact that the Chinese both bridge and challenge these paradigms further makes them an enigma to observers.

Despite the highly publicised anti-Chinese rhetoric in Zambia, local officials are very aware of the important role the Chinese play in-country. The Chinese influence is evident in the number of individuals present when stepping into any major government office, and extends beyond the official sphere into public life via daily vegetable markets and the recent proliferation of Chinese restaurants. The Zambian government's recent commitment to fund public school Chinese-language instruction is only one example of its long-run dedication to the bilateral relationship. Despite the temporary nature of many Chinese migrants, the community as a whole will only continue to grow.[1]

Migration to Zambia as an example of South–South migration does not fit the prevailing global model of low-skilled migration to high-income countries. Zambian labour immigration is largely characterised by educated foreigners offering sectoral expertise and management experience. Labour migrants enter the country not to occupy low-skilled jobs undesirable to 'first-tier' natives, but instead to manage and share expertise with a relatively uneducated Zambian populace.

9.3 Migration Policies and Regulations

There are many laws and policies on migration in Zambia. These laws are promulgated to protect, govern and address the plight and fate of migrants in Zambia. It is through the laws and policies that documentation such as visas and work permits are issued to identify and protect migrants and visitors. The most important legislation and policy governing migration in Zambia is the Government of the Republic of Zambia's (GRZ's) Immigration and Deportation Act No.18 of 2010 (Lupote, 2020).

The Immigration and Deportation Act No.18 of 2010 (replacing the original Act of 1965) regulates all migration matters, such as provisions for persons entering and leaving Zambia, immigration and residence permits, visas, border controls and prohibitions on human trafficking. The visa also stipulate how long one had to stay within the country. Any immigrant not possessing a visa is considered an illegal/undocumented immigrant. In 2017 the Refugee Act was promulgated. Through this Act the government ensures assistance and protection for any unaccompanied child, along with assistance to locate his/her parents and reunite with his/her family. When this is not possible, the child is accorded the same protection as any other minor deprived of his/her family. The Refugee Act of 2017 also provides for the refugee status determination procedure and addresses refugee law fundamentals, including the non-penalisation of irregular entry and presence in Zambia and the principle of non-refoulement.

Some immigrants entered Zambia for employment. For example, the University Teaching Hospital (UTH) and University of Zambia had expatriate doctors and lecturers, respectively. Also, some engineers engaged in road construction in Zambia were not Zambian. All these expatriates need to possess a work permit to work in Zambia. The GRZ (2010)[2] stipulates that work permits fall under section 28 of the Immigration and Deportation Act No.18 of 2010. Temporary employment permits (popularly known as travel documents in Zambia) are issued to business people intending to stay in Zambia for a period not exceeding 30 days. Section 28 of the Immigration and Deportation Act No.18 of 2010 provides for the issue of temporary employment permits.

Article 11 of the Constitution of Zambia Act No.1 of 2016 grants 'every person in Zambia' the right to life, liberty, security of person, protection of law, freedom of conscience, expression, assembly, movement and association as well as other rights and proceeds to provide that the exercise of these rights is irrespective of 'place of origin, race, political opinion, colour, creed, sex or marital status'. Yet there is Zambian jurisprudence that the rights enshrined in the Constitution do not extend to refugees.

The Constitution of Zambia Act No.1 of 2016 stipulates that Zambians and those in the diaspora are free to apply for dual citizenship. The same Act, according to 'Government decentralises issuance' (2016), stipulates that foreigners in Zambia could apply for dual citizenship in Zambia. The foreigners referred to here are the documented immigrants. Mwenya (2016) contend that the Act pleases many foreigners in Zambia and Zambians in the diaspora.

Among the immigrants in Zambia are the asylum seekers, particularly refugees. According to Chinyemba (2017), Zambia is part of the 1951 United Nations (UN) Refugee Convention and its 1967 Protocol that granted refugees the right to education, health, movement, integration and acquisition of travel documents. Apart from this international legislation and framework, Zambia also has its own legislation and framework that deal with refugees. In fact, for one to be considered a refugee in Zambia, Chitipula pointed out that he or she has to be subjected to Refugee Status Determination (RSD) procedures which the United Nations High Commissioner for Refugees (UNHCR) handled until 1993 when the Ministry of Home Affairs took over. Refugees are screened and admitted to refugee camps or settlements. 'The Zambian

Refugees Control Act of 1970 stipulates that refugees should reside in refugee settlements and camps' (Chinyemba, 2017). This contradicted the freedom of movement of refugees as stated in the 1967 protocol of the UN Refugee Convention. Under the Refugees Control Act of 1970, all refugees must live in an area designated by the Zambia government unless they receive special permission to remain outside. Section 16 of the Act allows an authorised officer to arrest a refugee without a warrant if they are 'reasonably suspected' of attempting to commit, or committing, an offence against the Refugee Control Act. Section 15 of the Act provides that breaches of the Act shall be punished with a period not exceeding three months imprisonment, in practice these periods are far longer. The Immigration and Deportation Act of 2010 states that asylum seekers should be granted temporary permission to reside in Zambia and this is to be renewed after it expires. Furthermore, the Act stipulated that it was an offence to reside outside a refugee camp or settlement, without permission. Chinyemba (2017), however, argued that according to the RSD procedures, refugees may live in urban areas and outside the refugee camp if they have employment, are investors, have family living in urban areas or suffer from health problems that are not treatable in the districts or are at risk of serious harm in the camps. Such refugees are to be given permission in the form of electronic refugee cards that expired, and had to be renewed after three years. The UNHCR work with the Zambian government and other stakeholders to ensure that refugees access services like education as well as the right to employment.

9.4 Migration Institutions and Governance

Zambia has several administrative and bureaucratic structures which were established to deal with matters affecting immigrants, in the government and NGO sectors. The government organisations are under different ministries and departments. The government departments and NGOs play different roles in the plight and fate of immigrants in Zambia.

9.4.1 The Zambian Immigration Department

The main governmental institution responsible for migration policy in Zambia is the Department of Immigration, established in 1965. The department, which falls under the Ministry of Home Affairs and has its headquarters in Lusaka and its regional offices in all Zambia's provinces, aims to regulate the entry and exit of people as well as their stay in the country in order to effectively increase internal security and sustainable socio-economic development.

In the Impala Documents (2015), it was noted that the department of immigration had its headquarters in Lusaka City. The department also had provincial headquarters and regional offices in the ten provinces of the country. Immigration offices are found at entry and exit points of the country such as the airports and border posts to monitor movements into and out of

Zambia. The immigration department undertakes many responsibilities aimed at immigration, emigration and also the plight of immigrants (documented and undocumented) in Zambia. According to Impala Documents (2015), the immigration department is responsible for facilitating and regulating the movement of people entering and leaving the country, and controlling the stay of immigrants and visitors in Zambia.

The department is responsible for issuing documentation to immigrants and dealing with undocumented immigrants in Zambia. According to GRZ (2010), the immigration department issued documentation to immigrants in accordance with the provisions of the Immigration and Deportation Act No.18 of 2010. The immigration department issue work permits, temporary work permits and visas.

The other duty of the immigration department is to deal with undocumented immigrants as well as overstaying visitors in the country. Occasionally, the immigration officers carry out patrols in residential areas to identify undocumented immigrants, and they work with the police to arrest, detain and prosecute them. In Lusaka City, the immigration officers carry out patrols especially in the informal settlements where undocumented immigrants seek refuge. According to Lupote (2020), the immigration department carried out a patrol and clean up exercise in March 2015 in Mutendere, Kalingalinga and Garden (informal settlements), where 19 undocumented immigrants were arrested. Simengwa (2014) pointed out that the immigration officers, in conjunction with the police and Registrar of Societies, arrested about 200 undocumented immigrants in Lusaka City – mainly from Burundi and Rwanda in 2014. The immigration department in January 2016 also arrested undocumented immigrants in Lusaka City after an alert from a local person (Chinyemba, 2017).

At the borders, the immigration officers searched vehicles entering the country as well as the documentation of persons intending to enter or leave the country. The immigration officers also mounted roadblocks on some roads – especially the four main highways – to check vehicles and passengers' documentation. The immigration department work hand in hand with the Zambia Police when it comes to the arrest and detention of undocumented and overstaying immigrants. Therefore, the Zambia Police is one of the important administrative and bureaucratic structures dealing with immigration and immigrants in Zambia.

The Ministry of Gender, the Ministry of Labour and Social Security, and the Central Statistical Office are also engaged in addressing the country's migration issues. On the one hand, the Ministry of Gender is mandated to promote gender equality by encouraging the development of national policies adequately responding to the different circumstances affecting migrant men and women. On the other hand, the responsibility for the formulation and administration of labour and employment policies – both in terms of labour migration and not – is entrusted to the Ministry of Labour and Social Security. The Ministry of Youth, Sports and Child Development addresses the challenges faced by the youth in terms of unemployment, poverty and vulnerability. Lastly, the Central Statistical Office, operating under the Census and Statistics Act, collects migration data to be included in the national data collection systems so as to make them available for policy development and planning (Migrants/Refugees, 2021).

The Zambia Police are committed to providing high quality service through applying the law fairly and firmly to everyone in the community. The national headquarters of the Zambia Police are found in Lusaka City, and provincial headquarters are found in all the ten provinces of Zambia. Furthermore, each district has a district police station, and police stations and posts are found in the settlement areas. The Zambia Police assist the immigration department in arresting and detaining undocumented immigrants. Once undocumented immigrants are identified, the Zambia Police come in to arrest, detain and prosecute them. Prosecution leads to removal or deportation, or even a jail sentence. The Zambia Police also show efforts in protecting non-Zambians in Lusaka City. This was noticed when the Zambia police officers were deployed to calm the riots that broke out from 19 to 21 April 2016 in Lusaka City's informal settlements such as George, Matero, Mandevu and Zingalume, where foreigners' shops were looted following a rumour that foreigners, particularly Rwandese, were involved in the ritual murders that swept through the informal settlements. The officers helped calm the situation, and the police posts and stations hosted the immigrants who sought protection to escape the riots (Chinyemba, 2017). The Zambia Police also closely guarded the St. Ignatius Parish in Lusaka City where some immigrants sought refuge.

Apart from the Zambia Police, the Drug Enforcement Commission (DEC) also interfaces with immigrants and targets those persons involved in smuggling drugs into and out of Zambia and those engaged in money laundering and document forgery. The Commission was established in 1989 under the statutory instrument No.87 of 1989. Usually, the DEC's concern is drug trafficking and money laundering. Drugs are sold within Zambia and smuggled into and out of the country. The local people and immigrants (both documented and undocumented) are engaged in the selling of drugs. The DEC works hand in hand with the immigration department, local people and the Zambia Police in identifying the people in possession of drugs, fake documentation, laundered money and obscene material such as pornography materials.

The department's national headquarters is in Lusaka City and it is subdivided into two: the national registration division and the passport and citizenship division. The national registration division issue NRCs to Zambians who are 16 years old and above so that they are identified as citizens. Non-Zambians/immigrants are granted NRCs once they are issued citizenship. To enhance its functioning, the department decentralised the issuance of NRCs from districts to sub-district offices ('Government decentralises issuance', 2014). The passport and citizenship division grants citizenship to non-Zambians/immigrants who wish to settle permanently in Zambia. Foreign nationals apply for, and acquire, citizenship and permanent stay in Zambia from the passport and citizenship division.

In addition to the government organisations, there are NGOs that deal with immigration and immigrants (both documented and undocumented) in Zambia. Such organisations are both local (Zambian) and international. Non-governmental organisations concerning migrants that operate in Zambia include the International Organization for Migration (IOM), the International Labour Organization (ILO), the UNHCR, the Organization of African Unity (OAU) and Doctors Without Borders. The European Union donated 1.6 million euros to various government initiatives and programmes to combat human trafficking which were implemented through the International Organisation for Migration (IOM), the International Labour Organisation (ILO) as well as the United Nations Children's Fund (UNICEF).

Apart from international NGOs, local NGOs such as faith-based organisations (FBOs) are active in the plight of immigrants, especially refugees. Different churches such as the Catholic Church play a vital role in ensuring that refugees have a safe stay in Zambia. FBOs, specifically churches, are mainly concerned with the vulnerable in society and concentrate their humanitarian work on migrants and refugees. Churches, especially the Catholic and Anglican churches, make donations in the form of clothes and food to the refugees in the refugee camps in Zambia. Also, the churches in Zambia protect refugees and host them when the need arises. Apart from accommodating and providing for refugees, the churches also empower the refugees through teaching them skills such as tailoring, gardening and carpentry. The skills enable refugees to be self-sufficient once integrated in society. The Catholic Commission for Development (CCD) (Caritas Zambia) plays a major role, as does the Jesuit Refugee Services (JRS) in Zambia. In 1994 the Zambia Episcopal Conference delegated the JRS through the CCD to carry out the Church's response to the issues of refugees in the country. The JRS started an Urban Refugee Project in 1997 in Lusaka. The JRS has always striven, through the Urban Refugee Project, to overcome the xenophobia which occasionally arises among Zambians by providing programmes which allow refugees and citizens to experience positive interaction and treat each other according to gospel values (Migrants/Refugees, 2021).

9.5 Summary

Since the demise of colonialism, Lusaka has maintained its profile as an attraction centre despite the fact that the fortunes from the copper mines have been fluctuating. Its favourable political climate, coupled with the friendly atmosphere of the people, has been a major attraction for those seeking to invest or stay in the city. Zambia, like other countries in the region, is a signatory member to both regional and international migration protocols, which it observes. It has both policy and organisational structures that facilitate the movement of people within and between borders. However, despite having such institutional frameworks in place, the country does not have freedom of movement for undocumented immigrants and refugees who are always kept in designated refugee camps outside the city.

Notes

1 https://www.migrationpolicy.org/article/following-money-chinese-labor-migration-zambia
2 Government of Zambia.

Bibliography

Awumbila, M. 2017. Drivers of Migration and Urbanization in Africa: Key Trends and Issues. *Proceedings of United Nations Expert Group Meeting on Sustainable Cities, Human Mobility and International Migration*, 7–8 September. New York, United States, https://pdfs.semanticscholar.org/36d3/22465721e1ae7ae4c596e287fe4a2ad414a1.pdf

Centre for Affordable Housing in Africa. 2017. 2017 Year Book. Housing Finance in Africa. A Review of Some of Africa's Housing Finance Markets. http://housingfinanceafrica.org/app/uploads/V18-Zambia-profile-final-18-Nov-2019.pdf

Chigavazira, B., Phillime, F., Kayula-Lesa, G., Shindondola-Mote, H., Frye, I., Nhampossa, J., Kaulem, J., Ramotso, M., Kafunda, M. & Mambea, S. 2012. Access to Socio-Economic Rights for Non-Nationals in the Southern African Development Community. *SPII and Osisa*.

Chinyemba, J. 2017. *Undocumented Immigration in Zambia: A Case Study of Lusaka City*. Pretoria: University of South Africa.

Coderre-Proulx, M., Campbell, B. & Mandé, I. 2016. *International Migrant Workers in the Mining Sector*. Geneva, Switzerland: International Labour Office.

Collins, J. 1969. *Lusaka: The Myth of the Garden City*. University of Zambia, Institute for Social Research.

Fargues, P. 2017. International Migration and Education—A Web of Mutual Causation. In *Think Piece Prepared for the 2019 Global Education Monitoring Report Consultation*. Paris: UNESCO.

Gann, L. H. 1964. *The Growth of a Plural Society: Social, Economic and Political Aspects of Northern Rhodesian Development 1890–1953, with Special Reference to the Problem of Race Relations*. University of Oxford.

Government Republic of Zambia. 2010. *Immigration and Deportation Act*. Lusaka: Government Printers.

Hall, R. 1965. Kaunda-Founder of Zambia.

IOM. 2019. Migration in Zambia: A Country Profile 2019. Https://Www.Zambiaimmigration.Gov.Zm/Wp-Content/Uploads/2021/05/Zambia-Migration-Profile-2019.Pdf

Lupote, I. S. 2020. *The Role of the Media in Curbing Unregistered Immigrants in Zambia*. Cavendish University.

Migrants/Refugees. 2021. *Migration Profile, Republic of Zambia*. Vatican City: Migrants & Refugees Section | Integral Human Development, https://migrants-refugees.va/country-profile/zambia/

Monu, J. V., Kopakopa, D. & Tajanovic, M. 2012. 'ISS outreach SubSaharan Africa: Zambia, 2011'. *Skeletal Radiology*, 41(12), 1493–1494.

Mulenga, C. L. 2003. Lusaka, Zambia. *Case Study for the United Nations Global Report on Human Settlements*.

Mwakikagile, G. 2010. *Zambia: Life in an African Country*. New Africa Press.

Mwenya, G. (2016, January 11). Zambians in USA applaud President Lungu over dual citizenship. *Zambia Reports*. https://zambiareports.com/2016/01/11/zambians-in-usa-applaud-pres-lunguover-dual-citizenship/

Nthanda, N. 2016. Xenophobic Attacks Erupt in Lusaka. *Iol*. https://www.iol.co.za/news/africa/xenophobic-attacks-erupt-in-lusaka-2011931

Pasteur, D. 1982. *The Management of Squatter Upgrading in Lusaka, Phase 2: The Transition to Maintenance and Further Development*. Development Administration Group, Institute of Local Government Studies.

Seshamani, V. 2002. *Zambia Poverty Reduction Strategy Paper: 2002–2004'*. Lusaka: Unpublished.

Simengwa, C. 2014, August 6. Zambia: Is Zambia a safe haven for illegal immigrants? *Times of Zambia*. www.times.co.zm/?p=29428

UN DESA. 2004. *Proceedings of the Third Coordination Meeting on International Migration*. New York. www.un.org/en/development/desa/population/migration/events/coordination/4/docs/Report_Third_Coordinationmeeting.pdf

Williams, G. J. 1984. *The Peugeot Guide to Lusaka*. Zambia Geographical Association.

10 In the Shadows of the Sunshine City of Zimbabwe

Harare

10.1 Introduction

What we know as Zimbabwe today began its formation when refugees fleeing Zulu violence or migrating from Boer moved to the land from the south in the 1830s. Around the same time, Europeans began exploring and expressing interest in the region. Zimbabwe was declared a British colony in 1889, bringing an influx of European settlers. Zimbabwe became an independent country under the rule of the white minority in 1965, causing civil and international outrage. Zimbabwe gained full independence in 1980.

Harare, formerly Salisbury, capital of Zimbabwe, lies in the north-eastern part of the country. The city was founded in 1890 at the spot where the British South Africa Company's Pioneer Column halted its march into Mashonaland; it was named for Lord Salisbury, then British prime minister. The name Harare is derived from that of the outcast Chief Neharawe, who, with his people, occupied the *kopje* (the hill at the foot of which the commercial area grew) at the time the Pioneer Column arrived and seized the land. The city was created a municipality in 1897 and developed after the arrival of the railway (1899) from the port of Beira, Mozambique, becoming a market and mining centre. It was chartered as a city in 1935. Industrialisation during and after the Second World War led to an influx of population. Salisbury was the capital of the colony of Southern Rhodesia, of the short-lived Federation of Rhodesia and Nyasaland (1953–1963) and of Rhodesia during the period of the unilateral declaration of independence (1965–1979). It was retained as capital by the new government of independent Zimbabwe (1980) and renamed Harare.

Harare lies at an elevation of 4,865 feet (1,483 m) and has a temperate climate. It is a hub of rail, road and air transport (the airport at nearby Kentucky handles international traffic) and is the centre of Zimbabwe's industry and commerce. It is also the main distribution point for the agricultural produce of the surrounding area, especially its Virginia tobacco. There are also important gold mines in the vicinity. Greater Harare includes residential Highlands and the industrial suburbs of Southerton, Graniteside and Workington. The most populous of the adjoining townships is Highfields.

The city is modern and well planned, with multistoried buildings and tree-lined avenues. It is the site of Anglican and Roman Catholic cathedrals,

DOI: 10.4324/9781003184508-10

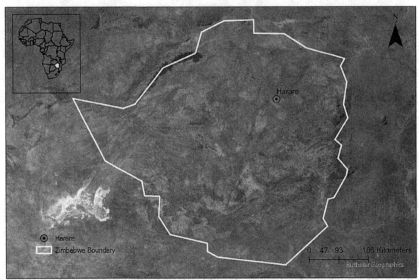

Figure 10.1 Map of Zimbabwe Showing the Position of Harare

a Dutch Reformed church, the Queen Victoria Memorial Library and Museum, the National Archives, the University of Zimbabwe (opened 1957) and the Rhodes National Gallery. Despite over a decade of neglect, the city's infrastructure and human capital still compares favourably with cities in other parts of Africa and those in Latin America. It remains to be seen whether the current government can entice its young, diverse and well-educated Zimbabwean diaspora, numbering some 4–7 million people, to invest in the economy, let alone consider returning. It hosts refugees mostly arriving from the Great Lake Regions. Thus, this chapter looks at the experiences of migrants, based on the qualitative study conducted in Harare (Figure 10.1).

10.2 Migrants' Profile in Harare

Intensive emigration from Zimbabwe in the last decade masks the fact that for over a century the country was both a migrant sending and receiving country (Maphosa, 2005). Before independence in 1980, immigrants came to Zimbabwe from Zambia, Malawi, Asia, Mozambique and Europe attracted by economic prospects in agriculture and mining (Tevera and Zinyama, 2002). This is because during the 1980s, after gaining its independence, Zimbabwe became an attractive place for foreign nationals as the newly elected leader of the nation, Robert Mugabe, preached hope and reconciliation following a bitter and protracted war for independence. Zimbabwe became the fourth most industrialised country in Africa south of the Sahara, with a middle-income status supported by a diversified economy (Nhema, 2002; Sachikonye, 2002). It has continued playing the dual role as both a receiver of migrant labourers from its neighbours and as a supplier of migrant labour to South Africa. Zimbabwe's economic prosperity was short-lived as the country started experiencing an economic decline in the late 1990s. Due to the economic challenges the country went through during the past decades, most of the immigrants have since moved to other Southern African countries such as South Africa and Botswana, but there are still some who have remained in Zimbabwe.

Interviews with government officials reveal that beyond being a receiver and sender of migrants, Zimbabwe is a transit route to other countries with better economic prospects and political stability than Zimbabwe's. Due to its proximity to South Africa, Zimbabwe has been a transit area for migrants arriving from countries in the Horn and Central Africa such as Burundi, the Democratic Republic of the Congo (DRC), Ethiopia, Somalia, Eritrea, Tanzania and others en route to South Africa (IOM, 2021). Being used merely as a conduit by migrant labourers from Malawi and Mozambique en route to South Africa, sometimes they would work in Zimbabwe for a while to earn enough to finance their journey southward and then move on.

According to the 2019 United Nations Department of Economic and Social Affairs (UN DESA) report, there were an estimated 411,300 international migrants living in Zimbabwe, and 87% of them were coming from five countries: Mozambique (160,000), Malawi (98,383), Zambia (26,909), the United Kingdom (15,561) and South Africa (11,571). Data from the Department of Immigration Control shed light on the number of Temporary Employment Permits (TEPs) issued to foreign nationals. Between 2010 and 2016, the Government of Zimbabwe issued 18,436 TEPs to foreign nationals from 74 countries, but most of them were Chinese nationals (11,272), accounting for 71%, and South Africans (1,859), followed by others coming from India and Zambia (IOM, 2021).

Zimbabwe hosts refugees mostly arriving from the Great Lake Regions. Zimbabwe adopts the camp model to handle refugees. There are two refugee camps in Zimbabwe: the Tongogara refugee camp in the Manicaland Province and the Harare refugee camp. According to the 2021 UNHCR report, there were 22,600 people of concern in Zimbabwe, more than 53% of them coming from the Democratic Republic of the Congo (12,020), Mozambique (3157), Rwanda (850), Burundi (964) and others (609) (UNHCR, 2021). Nationals from the Great Lake regions make up more than 85% of asylum seekers and refugees in Zimbabwe, which is also home to a small number of refugees from Somalia, Eritrea and Ethiopia.

Being a citizen of a state gives a person the right to access and exercise rights such as education, health and work. Being a stateless person, meaning a person who is not recognised by any state as its national, sometimes deprives that person the right to access basic services, economic opportunities and safety. The UNHCR Assistant High Commissioner, Gillian Triggs, acknowledges the Government of Zimbabwe for carrying out livelihoods projects to support the self-reliance of the people living in the refugees' camps. She stated that:

> Access to livelihoods opportunities, skills development, education, health, safety and security including peaceful co-existence with the host community are contributing to a favourable protection environment. Besides, access to cash to facilitate their daily lives is fundamental for refugees to live in dignity. I'm pleased to see that the Government, together with partners and UNHCR in the country, are working on this.
> (UNHCR, 2022)

In as far as access to housing migrants is concerned, one thing that is clear about migrants is that they find their initial phases of settling bewildering because they lack the requisite information about the host nation. In this study Harare scored high on challenges that are faced by foreign migrants to access housing. For forced migrants, Zimbabwe has an encampment policy requiring all refugees to stay in refugee camps (Chikanda and Crush, 2016). The model of refugee settlement developed and implemented by the Government of Zimbabwe in the 1980s and early 1990s was informed by the large-scale entry of Mozambican refugees into the country. The settlement of refugees in camps was the favoured option as it allowed the government to transfer the burden of day-to-day care of refugees to the UNHCR. Even though Zimbabwe's refugee policy allows for the integration of refugees into towns and villages, the government prefers to have refugees settle in camps where they can receive support from the UNHCR and other humanitarian groups (Chikanda and Crush, 2016). In support of the findings from this study, the UNHCR (2018) states that access to shelter remains a challenge due to funding constraints and increasing refugee population in the Tongogara refugee camp. The state of housing at the Tongogara camp is of major concern in that, while the Zimbabwean government has provided land, it has been appealing to donors for assistance with building materials, with limited success. Consequently, houses have been built with self-made mud bricks, which are of such poor quality that in early 2017, nearly 300 refugee homes were destroyed by heavy rains and storms (ClubofMozambique, 2017). The UNHCR is exploring environmentally friendly and low-cost shelter models. The UNHCR distributes Core Relief Items, including blankets, jerry cans, mosquito nets, soap, sanitary pads, kitchen sets and solar lamps, to person of concern in the camp. Adequate housing transcends beyond the physical structure to access to services such as energy, water and sanitation, and so on. The UNHCR implements WASH in partnership with GOAL Ireland. Activities are aimed at maintaining supply of potable water and increasing knowledge on WASH practices. Access to energy remains very low, with a monthly provision of 270 metric tonnes of firewood. The UNHCR is exploring to introduce more innovative/sustainable energy solutions (UNHCR, 2018). However, the situation is difference for expatriates, and other migrants who are in the country legally for the purposes of work, business and education.

The adoption of a highly spatialised policy of placing the refugees in camps was also meant to ensure that the increasing number of refugees would not put additional strain on Zimbabwe's job market, which was under severe pressure with the introduction of the Economic Structural Adjustment Programme (ESAP) in 1991(Chikanda and Crush, 2016). However, this did not spare the Zimbabwean labour market from the economic meltdown. The people of Zimbabwe have long experienced poverty, and it came to a head in the early 2000s when a state of disaster was declared due to food shortages. Harare's economy is the persistent emigration of highly educated and skilled residents to the United Kingdom, Australia, Canada, the Republic of Ireland and New Zealand, largely due to the economic downturn and political unrest. The city's brain drain, almost unprecedented compared to other emerging markets, has led to the decline of a local entrepreneurial class, an overstretched and declining middle class and a dearth of employment opportunities outside the informal and public sector. In addition, the city's working-class residents are increasingly moving to nearby South Africa and Botswana, though they are readily replaced by less well-off rural migrants. Economic opportunities are difficult to come by for the majority of forced migrants. Some of the labour immigrants use illegal means to get employment, while others come through the formal channel. The Ministry of Labour and Social Services issues work permits to those labour immigrants who come to work legally in Zimbabwe. Most of these registered foreign workers are members of international companies, NGOs and volunteer organisations. Refugees do not have formal access to the labour market and are therefore compelled to work in the informal sector, often working under duress, or in jobs presenting special hazards and risks. In order to reside in urban areas, refugees have to demonstrate that they have the wherewithal to fend for themselves, yet they are not allowed to seek formal wage earning employment. Nonetheless, refugees with resources to run private businesses are authorised to do so, and those who are qualified in professions with limited human resources, such as health services, may be allowed to work.

Mangezvo (2018) observed that Nigerian migrants represent the largest proportion of foreign entrepreneurs operating informal businesses in Harare. Migrants can be broken down to fluid ethnic groups from one country. However, these ethnic identities are de-emphasised, and the common identity as 'Malawian, Mozambican, Zambia, Chinese, Indian, Somalian, Nigerian etc' is the one that is preferred. These migrants chiefly come for business opportunities which they say are endless in Zimbabwe. Even those who view Zimbabwe as a transit point to other destinations end up attracted by these opportunities. In Harare, migrants, particularly Nigerians, have taken over the downtown area, putting up shops retailing goods and services ranging from motor spares, electricals, cell phones, clothing, cosmetics and internet cafes. However, in this same locale, migrants are often regarded as 'aliens and unwanted intruders' who have come to grab opportunities from the autochthons, and this further compounds the precarity of some migrants in Harare.

Migrants are generally considered a vulnerable population, with migrant women and children (new-born and young children) being particularly so. Access to health care for this population remains a challenge (Meyer-Weitz et al., 2018). The country's health sector faces numerous challenges: a shortage of skilled professionals and health care staff; an eroded infrastructure with ill-equipped hospitals, many lacking functional laundry machines, kitchen equipment and boilers; and a lack of essential medicines and commodities. The system breakdown has been exacerbated by humanitarian crises such as the cholera and measles epidemics between 2008 and 2010, by poor maternal and child health services and by consistently falling but nevertheless still-high numbers of people living with HIV. Interviews carried out with government officials for this study reveal that migrants face a higher degree of difficulties in accessing health care services. This is at the backdrop of the deteriorating Zimbabwe's health care services which coincided with a fall in demand for services, following the introduction of user fees. These fees, which are often applied in an ad hoc way and so vary from provider to provider, act as a barrier to basic health services for many of the most vulnerable people in Zimbabwe. Government policy is to provide free-of-charge health services for pregnant and lactating mothers, children under five and those aged 60 years and over, but the policy has proved to be difficult to implement. The situation is unattainable for refugees and asylum seekers who are confined in camps with limited access to economic or livelihood opportunities to support them to pay for health care services. In Zimbabwe, the public health system is the largest provider of health care services, complemented by Mission hospitals and health care delivered by non-governmental organisations (NGOs). In recent years, the economic decline and political instability have led to a reduction in health care budgets, affecting provision at all levels. In the past two decades, the country's poorest have suffered the most, with over 40% drop in health care coverage.[1] Currently, in the absence of substantial government financial support, user fees provide the main income for many health care facilities, enabling them to provide at least the minimum service. Considering the precarious condition of health care services for Zimbabwean nationals, migrants', especially forced migrants', access to the same health care services is inconceivable and dicey. The UNHCR (2016) notes that the clinic at the Tongogara refugee camp had attended to some 2,700 people in 2015, due to the collapse of Zimbabwe's national health care system, the cost for secondary and tertiary medical care increased. Against this background, it is probably unsurprising that the NGO Terre Des Hommes, which provides health and education support at the camp, has expressed difficulties in coping with increased demands for health services (ClubofMozambique, 2017). The clinic at Tongogara Refugee Camp provides 24-hour outpatient and referral services to refugees, asylum seekers and the host community. Screening, treatment and referrals to major government hospitals are provided free of charge. The clinic also provides reproductive health and HIV services (UNHCR, 2018). Triggs, the UNHCR Assistant High Commissioner, thanked the government for including refugees and asylum seekers in the national COVID-19 response and related national vaccination plans. Refugees and asylum seekers in the country have access to the health and education systems (UNHCR, 2022).

It is clear that apart from housing and health care services, commitment to education is one of the best forms of raising living standards for migrants and displaced population. The UNHCR supports early childhood, primary and secondary school education to persons of concern residing at the Tongogara refugee camp in Chipinge District (UNHCR, 2018). The UNHCR (2016) reports that in 2015, 1,600 children were enrolled for primary education at the government-run Tongogara primary school, while the teacher in charge laments the lack of teachers, classrooms and stationery (ClubofMozambique, 2017). Pupils who excelled academically used to be sent, by the UNHCR, to complete their secondary education at boarding schools outside the camp (Ghelli, 2016). With the budget not growing in line with continuous new arrivals in the camp, anecdotal evidence is that since 2016, the UNHCR has only been able to pay boarding school fees for pupils at a senior level of study (Taruvinga et al., 2021). Now high school level education is made available in public schools; however, there is a need to increase the number of teachers and classrooms (UNHCR, 2018). University-level education is only available for a limited number of refugee students due to insufficient funding for tertiary education and prohibitive costs that are beyond the reach of most refugees.

The asylum climate in Zimbabwe remains positive. Persons seeking asylum (primarily from the Great lakes region and the Horn of Africa) are afforded access to the country and the relevant procedures. Zimbabwe is a signatory to the 1951 Convention, with reservations in matters touching on employment and freedom of movement, as well as to the 1969 OAU Convention. These two major Conventions together with the Refugee Act adopted in 1983 provide adequate juridical framework for the UNHCR activities in the country. Restrictions on freedom of movement and employment, grounded in the country's reservations to Articles 17 and 26 of the 1951 Convention relating to the Status of Refugees, have meant that refugees are obliged to reside in the Tongogara refugee camp and are fully dependent on humanitarian assistance without any meaningful prospects for local integration (UNHCR and WFP, 2014). Detention is used for migration-related offences, including unlawful entry, employment without an official permission to work and exiting the refugee camp without authorisation. Notwithstanding the use of detention in these cases, about 14% of refugees and asylum seekers live in urban areas. As of 1 January 2015, there were 100 known registered refugees and asylum seekers who had been detained throughout 2014 across the country for migration-related offenses and were eventually released though the UNHCR's intervention. This said, the government has shown flexibility in allowing refugees to reside elsewhere, principally in Harare and Bulawayo, and take up employment in the informal sector and in issuing work permits for professionals (UNHCR and WFP, 2014).

In interviews with government officials and academic professionals, exploitation of migrants was indicated to be low. This could be primarily because migrants are either protected in camps or self-employed in urban areas like Harare or are employed formally with proper documentation which guarantees their right against exploitation. The exception is true for irregular migrants who enter the country through extra-legal means or are trafficked into the country. At the beginning of 2013, the government acceded to the Palermo Protocol to Prevent, Suppress and Punish Trafficking in Persons, especially Women and Children. Furthermore, on 3 January 2014, the government passed regulations to criminalise human trafficking through a Temporary Measures Act. The regulations are designed to give temporary legal effect in Zimbabwe to the country's obligations under the 'Palermo Protocol.' (UNHCR and WFP, 2014). Despite its efforts in combating trafficking, which included training of 264 detectives on anti-trafficking law and victim protection, and the incorporation of a module on the anti-trafficking law for new police recruits from 2019, Zimbabwe does not fully meet the minimum standards for the elimination of human trafficking. For example, the draft amendments to the 2014 trafficking in Persons Act to bring the law in line with international standards by 2020 has not yet been ratified, and the government failed to provide adequate funding to its NGO partners on which it relied to provide protection services to victims (US Department of State, 2020).

According to the 2020 United States Department of States Trafficking in Person Report (TPR), 'Zimbabwe is a source, transit, and destination country for men, women, and children trafficked for forced labour and sexual exploitation' (US Department of State, 2020). This report shows that children are being smuggled from Mozambique into Zimbabwe and are forced into street vending. Porous borders, economic hardship and opportunity deprivation have made Zimbabweans vulnerable to false promises of having a better life elsewhere. There are reports of Zimbabwean women being lured to China and the Middle East for work. Traffickers entice Zimbabwean men and women into hard labour conditions in agriculture, construction, information technology and hospitality, mostly working in neighbouring countries. Some of them subsequently become victims of forced labour and some women are also involved in sex trafficking. Many Zimbabwean adults and children get into South Africa, where traffickers exploit them in labour and sex trafficking (Migrants/Refugees, 2022).

Local integration, in particular options for naturalisation, has not been supported by the Government of Zimbabwe and is thus not a real option to address the protracted nature of asylum in Zimbabwe. This leaves resettlement as the only 'certain' durable solution today. Care is, however, taken to ensure that this solution does not become a magnet attracting individuals to Zimbabwe only for the purpose of resettlement (UNHCR and WFP, 2014).

The period from 1995 to 1999 marked a policy shift by the Zimbabwean government which had hitherto promoted the settlement of refugees in camps. Given that the country hosted its lowest ever refugee population during this period, refugees were no longer accommodated in refugee camps following the closure of all five camps in 1995. However, in 1999, the government reopened Tongogara camp in response to the growing influx of refugees from the Great Lakes region. This marked a reversion to the favoured encampment policy. Only in exceptional cases were refugees officially allowed to settle in urban areas, mainly Harare and in 1999, the UNHCR reported that some refugees were being relocated to Tongogara camp from urban areas (UNHCR, 2012). The local integration of refugees has not been viewed as a favoured policy strategy. At the camp, the refugees have access to small-scale agricultural opportunities, as well as microfinance in support of other income-generating ventures which gives them a certain level of self-sufficiency (UNHCR, 2000).

Generally, when migrants move to new territories they are regarded as aliens; and they become scapegoats when nation-states confront economic, political and social problems. They become targets of hostility from the autochthons and blamed for all the social, economic and political problems in the host nation. Zimbabwean bureaucrats suspect migrants of being drug dealers, thieves, fraudsters and con artists (Mangezvo, 2018). This means that Nigerian migrants have to devise new strategies of surviving such a toxic environment. Nigerian migrant entrepreneurs have learnt that the only resolution to disputes with their hosts is through forming an association, which plays a critical judicial function in advocating for their rights. Despite the documents they have to legitimise their stay, they remain subject to multiple forms of exactions from various rent-seeking individuals, maltreatment and unlawful arrests. Rather that exercising the 'politics of invisibility', which characterises most migrants in a foreign land (Whitehouse, 2012), they have opted for the 'politics of recognition' (Gutmann, 1994, Englund and Nyamnjoh, 2004) through the association, to defend their interests en masse – by collectively refusing to put up with abuse, being cheated and paying bribes.

It is a very difficult position for foreign migrants to advocate for their rights in this Zimbabwe. They are limited to do what they want, even through legal means. Taking a grand stand of 'politics of recognition' through launching of complains or reporting crime to institutions such as the Zimbabwe Anti-Corruption Commission (ZACC) does not go down well with some Zimbabweans, who think that this is a breach of their tacit agreement not to 'rock the boat' in the host country. This narrative concurs with Whitehouse's (2012) observations that strangers in a foreign land should have a stranger's code that obliges them to desist from any activity that might disturb the tenuous relationships between them and the hosts. This includes desisting from drawing unwanted scrutiny to the locals in exchange for permission for 'belonging' in the host country. By this, Whitehouse (2012) argues that strangers should remain quiet and keep their heads down, as exile knows no dignity.

Migrant associations foster social ties, which facilitate entrepreneurship among migrants (Putnam, 2000). Norris (1975) also pointed out that migrant associations provide newcomers with a basis of familiar and viable interactions and relationships on which they start to build their new lives 'elsewhere'. However, besides borrowing from the association, some have also borrowed from their Zimbabwean counterparts.

10.3 Migration Policies and the Regulatory Environment

In addition to its national legal framework governing migration, Zimbabwe is a signatory to the continental, regional and international legal framework regulating and protecting the interest of people on the move.

On the international front, Zimbabwe is a party to the international convention on the Protection of the Rights of All Migrant Workers and Members of their Families, the 1951 Geneva Convention relating to the Status of Refugees and the 1967 New York Protocol relating to the Status of Refugees (hereinafter jointly referred to as the 1951 Convention) in 1981, However, it has entered reservations to Article 17 (wage-earning employment), Article 23 (public relief), Article 24 (social security) and Article 26 (freedom of movement). Zimbabwe deposited the instrument of accession to the 1954 Convention relating to the Status of Stateless Persons (the 1954 Convention) in 1998, but it has not yet acceded to the 1961 Convention on the Reduction of Statelessness (the 1961 Convention). During the first cycle of the UPR in 2011, Zimbabwe accepted a recommendation to accede to the 1961 Convention. It is part of the 2000 Protocol to Prevent, Supress, and Punish Trafficking in Persons, especially Women and Children, the International Convention on the Elimination of All Forms of Racial Discrimination (ICERD), the Convention on the Elimination of All Forms of Discrimination against Women (CEDAW) and the International Convention on the Rights of the Child. These international agreements protect all human beings regardless of their nationality.

At the continental level, Zimbabwe ratified the African Union Convention for the Protection and Assistance of Internally Displaced Persons in Africa (Kampala Convention) and the 1969 OAU Convention Governing the Specific Aspects of Refugee Problems in Africa. The 1969 OAU Convention Governing the Specific Aspects of Refugees in Africa (the 1969 OAU Convention). Zimbabwe is a member of the Common Market for Eastern and Southern Africa (COMESA), whose treaty advocates for the suppression of obstacles to the free movement of people, as well as the recognition of the establishment and residence among member states.

At the regional level, Zimbabwe has a Bilateral Labour Agreement with South Africa for the facilitation of the recruitment for commercial farms in the Limpopo Province in South Africa. Zimbabwe is also a member of the Migration Dialogue for Southern Africa, which aims at promoting cooperation among SADC member states on migration-related issues, thus enhancing their capacity to manage migration within a regional context.

On the national front, the 1997 Immigration Act, amended in 2001, is administered under the Department of Immigration Control by the Minister of Home Affairs. The Act gives the Department authority to regulate both entry into and exit of nationals and foreign nationals from Zimbabwe. Upon the authority given to them by this Act, the Department of Immigration Control is in a position to supply information as to how many people have crossed the Zimbabwe borders, how many have been refused entry into Zimbabwe and how many have applied for either the refugee or asylum seeker status at any point in time and those who have been granted the status.

There has been a considerable evolution in the migration trajectory in Zimbabwe. This has prompted the government to respond by joining forces with the IOM to form a new national migration strategy that culminated after several workshops in the drafting of a National Migration Management and Diaspora Policy. The national migration strategy involves officials from the Migration and Development Unit of the Ministry of Economic Planning and Investment Promotion, the Ministry of Labour and Social Welfare, the Ministry of Justice (Chereni and Bongo, 2018), the Legal and Parliamentary Affairs, and IOM. The Draft policy focuses more on mitigating the challenges the country is facing with the large emigration and cross border movements. It aims at the retention and return of highly skilled Zimbabwean nationals and also to promote strategies aimed at opening new channels for legal migration of low and semiskilled workers (*ibid*). The Draft policy foresees the ratification of the Protocol to Prevent, Suppress and Punish Trafficking in Person especially Women and Children, and the Protocol against the Smuggling of Migrants by Land, Sea, and Air (*ibid*).

The legal framework on nationality in Zimbabwe also consists of Chapter 3 of the 2013 Constitution and the Citizenship of Zimbabwe Act No. 23 of 1984 (as amended by Act No. 7 of 1990, Act No. 12 of 2001, Act No. 22 of 2001, Act No. 23 of 2001, Act No. 1 of 2002 and Act No. 12 of 2003). It contains some important safeguards against statelessness; however, gaps remain, including the granting of nationality for children born to stateless parents. The 2013 Constitution states that the right to have access to basic health care services is given only to citizens and permanent residents. Refugees and other migrants are excluded. However, the law further remarks that any person living with a chronic disease, including migrants, also have the same right, and children below the age of 18 and the elderly have the right to health care services (Republic of Zimbabwe, 2013). Section 65 of the 2013 Constitution recognises the right to fair and safe labour practices and standards for everyone, citizen or non-citizen (*ibid*), and this automatically covers migrant workers. Section 64 further emphasises 'the right of everyone to choose and carry on any profession, trade or occupation' in Zimbabwe (Republic of Zimbabwe, 2013). This protects the employment rights of workers from undue restriction, unfair and unsafe treatment, and unjust discrimination. The law also provides migrant workers the possibility to form, join and participate in trade unions.

The 2013 (or new) Constitution contains provisions that could provide the basis to prevent and reduce statelessness in Zimbabwe. For instance, Article 39(3), which deals with revocation of citizenship, establishes that 'Zimbabwean citizenship must not be revoked under this section if the person would be rendered stateless'. Article 43 of the Constitution is equally important, since it opens the door to citizenship for persons born in Zimbabwe before the Constitution entered into force, where at least one of his or her parents were citizens of a country from the Southern African Development Community (SADC). This provision could potentially resolve the nationality situation of the vast majority of former farm or mine workers in the country.

Statelessness has been identified as a concern to the UNHCR mainly regarding the migrant population that came into Zimbabwe from neighbouring countries (Mozambique, Malawi and Zambia) as farm and mine workers during the colonial period. This population group was negatively affected by multiple changes in nationality laws in Zimbabwe, especially in the period between 1963 and 2003. It is estimated that, at the height of agricultural expansion, between 20% and 30% of up to 2.5 million farm workers in Zimbabwe were of foreign ancestry. Based on these numbers, and taking into account that some of these persons have already resolved their nationality issues, the UNHCR estimates that 300,000 persons may be at risk of statelessness. Despite the gravity of the situation, the government lacks information on the scope of the problem and offers limited protection to the affected persons.

Although the new Constitution provides wider protections for persons at risk of statelessness, gaps remain when considering the protections advanced by the 1961 Convention, to which the country has not acceded. For instance, while Section 38 provides that persons continually residing Zimbabwe for at least ten years and satisfying certain conditions are entitled to apply for citizenship. This has been highlighted in the interviews with government officials. The law does not entirely protect persons that are stateless or at risk of statelessness. Additionally, while the constitution provides for the right of citizenship through a person's mother or father, the NGO Lawyers for Human Rights reports that, in practice, a single mother registering a child in accordance to the law may at times face challenges due to patriarchal attitudes that persist. Further, the Constitution does not guarantee that a person living abroad for extended periods of time or those who fail to register can retain their nationality. Additionally, there is also no guarantee that a person shall not lose nationality if that loss would render the person stateless.

The 1984 Citizenship of Zimbabwe Act (as amended in 1990) provides the process through which one may register as a citizen of Zimbabwe. This is a restrictive piece of legislation, especially when looked at in light of the new Constitution. It provides, inter alia, that non-Zimbabwean children should have their parents registered before they can themselves be registered as citizens of Zimbabwe. The Citizenship of Zimbabwe Act does not contain safeguards with respect to loss of citizenship. It appears to allow renunciation without the possession of or a guarantee to acquire another nationality. It also provides for the loss of nationality for those who reside abroad for more than seven years. As such, this Act needs to be amended and aligned with the new Constitution. While negotiations have continued towards accession to the Convention, certain forces in the government have discouraged the accession arguing, inter alia, that Zimbabwe does not have a statelessness problem.

The Refugees Act (Act 13/1983, 22/2001, (s.4)) is administered by the Ministry of Labour and Social Services with financial and material support from the United Nations High Commissioner for Refugees. The Act makes provision for refugees to enable effect to be given within Zimbabwe to the Convention Relating to the Status of Refugees, done at Geneva on 28 July 1951, to the Protocol Relating to the Status of Refugees of 31 January 1967 and to the Convention Governing the Specific Aspects of Refugee Problems in Africa, done at country level. Currently, there are quite a few refugees from countries such as Burundi, the Democratic Republic of Congo, Rwanda and Somalia. An Inter-ministerial Committee sits every other Thursday of the week to determine the status of refugees and asylum seekers. This Act implies that citizens of other countries have the right to apply for refugee status and this has a bearing on the total population in a country. Since 2012 the Government of Zimbabwe, through the Office of the Registrar General, has been carrying out occasional missions to the Tongogara refugee camp to issue birth certificates to refugee children. In 2015, the government issued birth certificates to 147 refugee children. A total of 556 males and 438 females received new identity cards and an additional 92 males and 47 females renewed their ID cards.

Zimbabwe has not been having a comprehensive and coherent legal framework for implementing migration practices and this has been affecting the capacity of the government to manage migration issues. Currently, the country through the Ministry of Economic Planning and Investment Promotion is in the process of finalising the National Migration Management and Development Policy. The Policy was read in the Parliament for the first time and comments were made, which are still being addressed. The policy aims at mainstreaming migration issues into the national policy framework. Among others, it focuses on the following key strategic areas:

- **Labour migration**
  The policy aims at formalising labour migration for national socio-economic development.
- **Irregular migration/Informal cross border traders**
  The policy seeks to facilitate the safe and legal migration of Zimbabweans. It also aims at promoting and protecting human rights and the well-being of migrants.
- **Human trafficking and smuggling**
  The policy aims to curb human trafficking and smuggling through the development of legislation that directly addresses the issues. It will push for the ratification of the Parlemo Protocol and other related instruments.
- **Migration and health**
  The policy will seek to ensure that migrant populations have access to health services. In addition, all migrants need access information on sexual and gender-based violence. It will also address the issue of HIV and AIDS comprehensively within and amongst migrant communities.

10.4 Migration Institutions and Governance

The state is one of the main actors in managing and governing migration, through different departments. The Zimbabwe National Statistic Agency (ZIMSTAT) is responsible for the population census in Zimbabwe. The Ministry of Home Affairs and Cultural Heritage is in charge of the identification of all people living in Zimbabwe. The Ministry of Home Affairs is also responsible for maintaining public order and security, also because the Police Department is part of this Ministry. It also controls the entry and exit of people across the Zimbabwe borders and is responsible for the issuance of personal documents like passports, refugee permits, etc. Through the department of Immigration Control, it also collects statistics for people coming into the country and going out of the country. Sits on the inter-ministerial committee to determine refugee status of applicants. The Ministry of Foreign Affairs assists in the drafting of the National Migration Management and Development policy document and acts as a conduct between the Government of Zimbabwe, in-transit and host countries and as a contact or conduit between the migrants and the government through embassies. It also maintains diplomatic relations with regional and international bodies such as Southern Africa Development Community (SADC) and the United Nations. Refugee protection is governed by the Ministry of Public Service, Labour and Social Welfare. This ministry, through department of labour, sits on the inter-ministerial committee that issues work permits to non-Zimbabweans. It sits on the inter-ministerial committee responsible for determining refugee status of applicants and compiles data on expatriates working in the country. The Ministry of Labour and Social Welfare through department of Social Services collects, compiles and disseminates statistics on refugees and asylum-seekers. It sits on the inter-ministerial committee responsible for determining refugee status of applicants. It intends to carry out a study to determine the number of displaced persons in Zimbabwe. The National Plan of Action is the implementation tool of the Trafficking in Person Act which is grounded on prosecution, prevention, protection and partnership. The Zimbabwe Refugee Committee (ZRC) is the national body mandated under the Refugee Act to conduct refugee status determination (CCJPZ, 2017). The Ministry of Economic Planning and Investment Promotion recommends policies for the enforcement of measures to protect and promote the human rights and well-being of migrants. It also identifies and recommends community development programmes to address the root causes of economically induced migration and provide livelihood alternatives for potential migrants. It liaises closely with the IOM Zimbabwe Office in implementing technical co-operation in identified areas.

Faith-based organisations (FBO), like the Catholic Church in Zimbabwe, play a pivotal role in assisting migrants and refugees thanks to their humanitarian efforts. One can mirror the humanitarian activities of the Catholic Church in Zimbabwe through its projects, involving agencies like the Jesuit

Refugee Service (JRS), Caritas and the Catholic Relief Services (CRS). Through its pastoral programs, the JRS provides skill training to refugees in one of the biggest refugee camps in Zimbabwe, the Tongogara refugee camp, and also to local Zimbabweans. Training is in the fields of motor mechanics, carpentry, agriculture, computers, hairdressing, cosmetology and sewing. The JRS also provides teaching equipment, infrastructure and uniforms to the secondary school in Tongogara. The CRS provides orphans and vulnerable children with education opportunities, income generating activities and psychosocial support. The CRS, in collaboration with the World Food Programme (WFP), has provided 3,099 malnourished HIV and TB patients, pregnant and lactating mothers, and children under five with food (CRS, 2019). This organisation also offers programmes to help victims of sexual violence resume their normal life through counselling, medical assistance and legal support. After being hit hard by a drought that destroyed their crops and their source of livelihood, Caritas Zimbabwe, working together with Caritas Internationalis and WFP, distributed food to 300,000 drought affected persons. The Catholic Commission for Justice and Peace in Zimbabwe (CCJPZ), a commission of the Zimbabwe Catholic Bishops' Conference, assists the vulnerable in cases of human rights abuses, through sensitisation campaigns aimed at educating people on their rights in Zimbabwe (CCJPZ, 2017).

International institutions and other organisations play a key role in migration management and governance in Zimbabwe. The International Organisation for Migration (IOM) Zimbabwe is recognised by the Government of Zimbabwe as the principal international inter-governmental organisation addressing the entire spectrum of migration issues (Migrants/Refugees, 2022). The IOM Zimbabwe has an exemplary track record in the provision of technical assistance on migration management, emergency response, HIV interventions, health management, protection of migrants' rights and dissemination of safe migration information. It offers technical and financial assistance to institutions dealing with migration issues and conducts research and produces reports on migration issues related to labour policies. It also assists in the repatriation of Zimbabwean nationals.

The IOM Zimbabwe is currently working in collaboration with the Government of Zimbabwe on:

- Provision of technical and capacity building assistance to local authorities and relevant ministries to facilitate the transition of mobile, migrant and displaced populations from a point of vulnerability to the achievement of sustainable, development-oriented solutions. The IOM supports communities highly impacted by mobility, migration and natural disasters to become more resilient, which includes their ability to withstand the impact of hazards, shocks and stresses in future;
- Provision of technical and capacity building assistance to the Government of Zimbabwe to strengthen national capacities on migration management. In this regard, the IOM Zimbabwe works with the government

around the formulation of migration policies in order to harness and maximise the development potential of migration, which include the:

1. National Diaspora Policy;
2. Migrant Resource Centre in Beitbridge; and
3. National Migration Policy;

- Strengthening the Government of Zimbabwe's capacity to manage safe and legal migration of Zimbabweans and raising awareness and strengthening the capacity of migration management authorities on Integrated Border Management (IBM);
- Strengthening policy, regulatory and institutional frameworks and capacities to effectively counter trafficking in persons into, within and from Zimbabwe. In this regard, the IOM supports awareness campaigns in order to sensitise and raise awareness on counter trafficking, provides direct assistance to victims of trafficking and provides technical assistance on identification of traffickers and victims of trafficking;
- Strengthening the capacity of the Ministry of Health and Child Care to manage and coordinate various migration health challenges and promote the health of migrants; and
- Supporting assisted voluntary returns back to Zimbabwe and resettlement of refugees to various resettlement countries.

IOM Zimbabwe is part of the United Nations Country Team (UNCT) and the Humanitarian Country Team (HCT). The UNCT and HCT are the highest-level inter-agency coordination and decision-making bodies for UN support to the Government of Zimbabwe and humanitarian interventions, respectively. Accordingly, the IOM participates in the development of emergency response plans which identify, prioritise needs and ensure a coordinated response in emergency situations. Further, the IOM participates in the development and implementation of the Zimbabwe United Nations Development Assistance Framework (ZUNDAF), a strategic planning instrument jointly formulated by the UNCT and the Government of Zimbabwe to identify national humanitarian and development priorities.

The United Nations High commissioner for Refugees (UNHCR) also plays an essential role in migration-related issues in the country, by assisting refugees, asylum seekers and community members in the areas of education, health, food security and nutrition, protection, water sanitation, and hygiene, camp coordination and management, access to energy, and community empowerment and self-reliance (Migrants/Refugees, 2022). The UNHCR offers technical and financial assistance to the Ministry of Labour and Social Services. It also collects and compiles data on refugees and asylum seekers and sits on the Inter-ministerial committee responsible for determining refugee status of applicants. The UNHCR's main objective is to improve living conditions for refugees, enhance their access to livelihoods and skills-training opportunities, and increase their access to durable solutions. To attain these goals, the UNHCR implements camp-based income-generation projects, improves access to education and health facilities and provides help to obtain civil status documentation. It will also refurbish infrastructure and engage in advocacy and capacity-building related to status determination process, with a view to strengthening Zimbabwe's management of refugee claims. In its overall protection coordination role, the UNHCR ensures regular dialogue, assessment of needs, response planning and implementation by various humanitarian stakeholders and the government. It also plays an active role in the UN Country Team, the UN Humanitarian Country Team and the Security Management Team. The Office has been actively engaged in the UNDAF process with other key stakeholders. In addition, the UNHCR participates in the IASC Inter-Cluster Forum and leads the protection cluster, providing overall coordination for the sub-clusters and focal points related to IDPs, gender-based violence, child protection and human rights, as well as the rule of law.

A partnership between the World Food Programme (WFP) and the United Nations High Commissioner for Refugees (UNHCR) has helped 13,702 of them persevere and rebuild their lives. The WFP, UNHCR, Terre Des Hommes Italia and Goal International have provided approximately 400 refugee households with farming inputs to use on the small plots of land provided to them by the Government of Zimbabwe. In total, the TRC holds 25 hectares of irrigated land. Each household has been allotted 500 square metres to cultivate. Refugees are assisted with vegetable seed packs so they can grow an assortment of produce on this land each year. They primarily grow bananas, sugar beans and potatoes. Additionally, this land is used for poultry and pig farming projects in the camp. The WFP supports Tongogara Refugee Camp alongside the UNHCR, which is focused on providing shelter; water and sanitation; and educational supplies. The refugee agency is also committed to expanding and diversifying income-generating projects to increase the resilience of those living in the camp.

Other migration-related UN agencies in Zimbabwe include the United Nations Development Programme (UNDP), working in the area of creating sustainable growth to improve people's lives, the United Nations Children's Fund (UNICEF), assisting refugee children with access to education, the WFP which helps food-insecure people, including refugees, to meet their basic food and nutrition requirements, the International Committee of the Red Cross (ICRC) working in the areas of advocacy and promoting detainee welfare, including migrants, supporting health services and restoring family links separated by disaster. Mercy Corps provides assistance and relief to people affected by natural disasters, like Cyclone Idai, and GOAL, which is designed to build resilience and sustainable livelihood by improving health, water, sanitation and hygiene (WASH) and nutrition systems, supporting refugees and strengthening the value chains needed to foster long-term financial security.

Christian Care, created by the Zimbabwe Council of Churches, is tasked with improving the lives of the disadvantaged, including refugees, through disaster relief programs, integrated rural development programs, advocacy and income-generating projects. There is also the Legal Resources Foundation helping refugees with advocacy. Even though they do not specialise in providing legal aid for refugees directly, they could potentially advise on a case or help in finding a suitable lawyer or organisation to provide support.

10.5 Summary

Since the advent of independence, *the sunshine city* has been gradually losing its splendour as the centre of attraction in the region and abroad. The fast-track land reform programme worsened its woes as migration patterns in the country changed in favour of emigration. The country has well-structured immigration policies supported by functional organisational structures that facilitate movement of people. It is also a signatory member to both international and regional migration protocols. However, its immigration policies are not very liberal in the sense that illegal migrants and refugees do not have freedom of movement since they are confined to designated refugee camps. Illegal migrants are deported once arrested. The country is still a favourable destination to those looking for opportunities despite its poor economic performance in recent years.

Note

1 http://www.unicef.org/infobycountry/zimbabwe_56573.html

Bibliography

CCJPZ. 2017. *Zimbabwe*. [Online]. https://www.peaceinsight.org/en/organisations/the-catholic-commission-for-justice-and-peace-in-zimbabwe-ccjpz/ [Accessed 14/08/2022].

Chereni, A. & Bongo, P. P. 2018. *Migration in Zimbabwe: A Country Profile 2010–2016*. International Organization for Migration (IOM).

Chikanda, A. & Crush, J. 2016. The Geography of Refugee Flows to Zimbabwe. *African Geographical Review*, 35, 18–34.

Clubofmozambique. 2017. Hardship for Refugees from Mozambique at Zimbabwe's Tongogara Refugee Camp. https://clubofmozambique.com

CRS. 2019. *Zimbabwe*. [Online]. https://www.vaticannews.va/en/africa/news/2020-02/caritas-zimbabwe-feeds-300-000-drought-affected-persons.html [Accessed 14/08/2022].

Englund, H. & Nyamnjoh, F. B. 2004. *Rights and the Politics of Recognition in Africa*. Zed Books.

Ghelli, T. 2016 Potatoes, Pigs and Poultry: Changing the Game for Refugees. www.unhcr.org

Gutmann, A. 1994. Multiculturalism: Examining the Politics of Recognition.

IOM. 2021. *International Organisation for Migration (Iom) Zimbabwe National Country Strategy 2021–2024* [Online]. https://reliefweb.int/sites/reliefweb.int/files/resources/zim%20national%20strategy.pdf [Accessed 21/02/2022].

Mangezvo, P. L. 2018. Catechism, Commerce and Categories: Nigerian Male Migrant Traders in Harare. In Oliver Bakewell & Loren B. Landau (Eds.), *Forging African Communities*. Springer.

Maphosa, F. 2005. *The Impact of Remittances from Zimbabweans Working in South Africa on Rural Livelihoods in the Southern Districts of Zimbabwe*. Citeseer.

Meyer-Weitz, A., Oppong Asante, K. & Lukobeka, B. J. 2018. Healthcare Service Delivery to Refugee Children from the Democratic Republic of Congo Living in Durban, South Africa: A Caregivers' Perspective. *Bmc Medicine*, 16, 1–12.

Migrants/Refugees. 2022. *Migration Profile, Zimbabwe*. Migrants And Refugees-Integral Human Development [Online]. https://migrants-refugees.va/fr/blog/country-profile/zimbabwe/ [Accessed 14/08/2022].

Nhema, A. G. 2002. *Democracy in Zimbabwe: From Liberation to Liberalization*. University Of Zimbabwe Publications Office.

Norris, J. 1975. Functions of Ethnic Organizations. In H. Palmer (Ed.), *Functions of Ethnic Organizations* (pp. 165–176). Copp Clark Publishing.

Putnam, R. D. 2000. *Bowling Alone: The Collapse and Revival of American Community*. Simon And Schuster.

Republic of Zimbabwe. 2013. *Constitution of Zimbabwe Amendment (20)*. Harare. https://www.constituteproject.org/constitution/zimbabwe_2013.pdf

Sachikonye, L. M. 2002. Whither Zimbabwe? Crisis & Democratisation. *Review of African Political Economy*, 29, 13–20.

Taruvinga, R., Hölscher, D. & Lombard, A. 2021. A Critical Ethics of Care Perspective on Refugee Income Generation: Towards Sustainable Policy and Practice in Zimbabwe's Tongogara Camp. *Ethics and Social Welfare*, 15, 36–51.

Tevera, D. & Zinyama, L. 2002. *Zimbabweans Who Move: Perspectives on International Migration in Zimbabwe*. IDASA and Queens University.

UNHCR. 2000. *UNHCR Global Report 1999*. United Nations High Commission for Refugees (UNHCR).

UNHCR. 2012. *UNHCR Global Report 2011*. United Nations High Commission for Refugees (Unhcr).

UNHCR. 2016. UNHCR Operation in Zimbabwe Fact Sheet. www.Unhcr.org

UNHCR. 2018. *Zimbabwe* [Online]. http://reporting.unhcr.org/Node/10232 [Accessed 2018].

UNHCR. 2021. *Zimbabwe: Fact Sheet* [Online]. http://reporting.unhcr.org/Zimbabwe [Accessed 21/02/2022].

UNHCR. 2022. *UNHCR'S Protection Chief Concludes Three-Day Visit to Zimbabwe* [Online]. https://www.unhcr.org/afr/news/press/2022/1/61ee72634/unhcrs-protection-chief-concludes-three-day-visit-to-zimbabwe.html [Accessed 07/07/2022].

UNHCR & WFP. 2014. *Zimbabwe - UNHCR/WFP Joint Assessment Mission Report: Tongogara Refugee Camp, September 2014* [Online]. https://www.wfp.org/Publications/Zimbabwe-Unhcr-Wfp-Joint-Assessment-Mission-Tongogara-Refugee-Camp-September-2014 [Accessed 14/08/2022].

US Department of State. 2020. *2020 Trafficking in Persons Report: Zimbabwe* [Online]. https://www.state.gov/reports/2020-trafficking-in-persons-report/imbabwe/ [Accessed 14/08/2022].

Whitehouse, B. 2012. *Migrants and Strangers in an African City: Exile, Dignity, Belonging*. Indiana University Press.

11 Inclusion of Foreign Migrants in South African Cities
An Aside

11.1 Introduction

Cities in South Africa, under municipal authorities, are constitutionally assigned a primary role of providing basic services to communities, regardless of race, gender and origin. National or provincial governments are responsible for the primary needs of migrants, such as access to shelter, health care, education and economic opportunities; safety and security – including proper treatment in detention; and administrative justice. Despite this, under Section 153(a) of the Constitution, municipalities have a responsibility to 'structure and manage its administration, budgeting and planning processes to give priority to the basic needs of their communities and to promote the social and economic development of the community'.

Landau et al. (2011) stress that across the majority of South African cities, officials continue to react to migrants by implicitly denying their presence, excluding them from developmental plans or tacitly condoning discrimination throughout the government bureaucracy and police. Migrants are members of the community entitled to government resources and are potential resources for communities, but in many cases, government officials see them as an illegitimate drain on public resources. In some municipalities, there is a distinct sense that current residents or 'ratepayers' deserve to be privileged over new arrivals or temporary residents. In others, officials hold fast to the idea that migration worsens violent crime, disease and unemployment. Still others insist that matters related to migration and human mobility are exclusively the responsibility of national government. These perceptions place migrants outside of the local government constituency, preventing officials from adopting pragmatic policies to address their developmental impact and provide for their needs.

The success of a city depends on authorities' ability to develop and respond to a nuanced and dynamic understanding of their constituencies. Due to a range of factors, including poor data collection and analysis, few municipalities are able to do so. Indeed, one of the most fundamental challenges to local government in protecting the rights and welfare of migrants and other residents is how little municipalities know about the people living in their areas of jurisdiction. There is a lack of information about the urban poor generally, and even less about geographically mobile people.

DOI: 10.4324/9781003184508-11

Inclusion of Foreign Migrants in South African Cities 173

Failure to manage migration is yielding undesired consequences for all residents of South Africa. However, when properly managed, migration need not lead to conflict and tensions, but can help to provide much-needed skills and entrepreneurial energy, at the same time boosting regional trade and integration and helping facilitate post-conflict reconstruction in countries of origin. This chapter looks at the extent of inclusion of international migrants with a particular focus on three South African Cities – viz. Durban, Cape Town and Pretoria.

11.2 Inclusiveness in Spatial Integration

With regard to spatial integration, respondents were asked a number of questions that sought to establish the extent to which migrants integrated into the local community. These included the level of comfort living in local communities, whether migrant feared hostility from locals, whether migrants preferred to live only with other foreign migrants and the level of fear of xenophobic attacks. Figure 11.1 shows that with regard to the level of comfort of living in local communities/integration in local community, about 31% of the migrants indicated that their level of comfort varied from very low to low. Thirty per cent were relatively/moderately comfortable staying in local communities, while about 39% expressed a high to very high degree of comfort living in the local communities. The responses pertaining to the level of comfort taking residence in local communities did not vary by gender. Moreover, the overall result appears to be generally the same across all three metropolitan areas with the percentage of those that expressed a moderate to high level of comfort living in local communities being higher than those who expressed a low level of comfort.

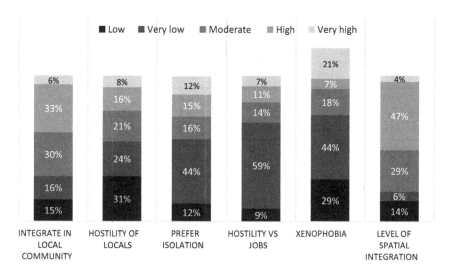

Figure 11.1 Spatial Integration in South African Communities
Source: Author (Fieldwork, 2022).

When asked how they would rate the level of spatial integration (acceptance of foreign migrants) of migrants into local communities, only 20% of the migrants rated it as low or very low, while 80% rated it as moderate to very high. These figures align with the observable reality that a lot of migrants, especially those from neighbouring SADC countries such as Zimbabwe, Malawi, Mozambique, eSwatini, Lesotho and Botswana, do indeed live among the local communities in township areas such as Mamelodi, Garankuwa and Soshanguve in Tshwane, Khayelitsha and Gugulethu in Cape Town, etc. Such migrants have assimilated very well into the local communities and they can speak one or two local languages. However, migrants from other African countries such as Nigeria, Somalia and the DRC tend to live in their own groupings and very few of them live in the townships. They probably account for much of the 31% that has a low to very low level of comfort living with the local communities. Some of these migrants cited their inability to speak any of the local languages as one of the reasons they found it difficult and uncomfortable to integrate into the local communities. Their zone of comfort is living among fellow foreign migrants. Hence, although there may be some fear of hostility from locals, foreign migrants have largely integrated into local communities (Tables 11.1 and 11.2, Figure 11.2).

The word cloud (Figure 11.3) captures the sentiments expressed by migrants when asked the extent to which they agreed that hostilities placed some limitation or restrictions on their movement in the city. As can be seen from the word cloud there is a very broad spectrum of responses. Some migrants do feel that they are limited or restricted in their movement.

Table 11.1 Level of Integration and Hostility towards Foreign Migrants

Spatial Integration	Integrate in Local Community (%)	Hostility of Locals (%)	Prefer Isolation (%)	Hostility vs Jobs (%)	Xenophobia (%)	Level of Spatial Integration (%)
low	31	55	56	68	36	20
Moderate	30	21	16	14	18	29
High	39	24	27	18	46	51

Source: Author (fieldwork, 2022).

Table 11.2 Rating Comfort Levels across Municipalities

Comfort level in Local Communities	Low (%)	Moderate (%)	High (%)
Cape Town	7	7	17
Tshwane	13	15	17
eThekwini	11	8	5
Total	31	30	39

Source: Author (fieldwork, 2022).

Inclusion of Foreign Migrants in South African Cities   175

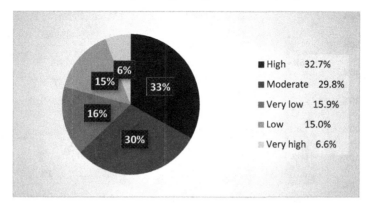

Figure 11.2 Level of Comfort in Local Residences
Source: Author, (Fieldwork, 2022).

Figure 11.3 Restrictions on Movement

Others indicated that they avoid visiting high risk areas or moving at night. Yet others felt that language barriers limited them from visiting certain areas since they cannot speak local languages (Figure 11.4).

The majority of respondents expressed sentiments that the city's efforts towards integration were very insignificant. This response was highlighted across the metropolitans. The cities need to consider the needs of about 20–30% of their populations to function more efficiently.

Table 11.3 shows how migrants rate the level of restriction or limitation on their movement within the city by province. It is noteworthy that a relatively

176 Inclusion of Foreign Migrants in South African Cities

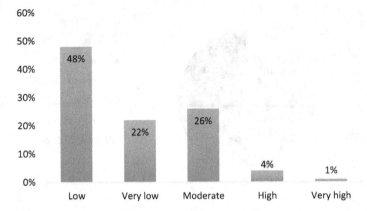

Figure 11.4 Rating Cities on Interventions for Improving Community Integration
Source: Author, (Fieldwork, 2022).

Table 11.3 Limitations or Restrictions on Your Movement in the City

Metropolitan	Race	Limitation or Restrictions on Your Movement in the City			
		High (%)	Low (%)	Moderate (%)	Total (%)
Cape Town	Asian	0.2	1.9	1.0	3.1
	Black African	1.0	15.2	3.1	19.3
	Coloured	0.4	1.1	0.2	1.7
	Indian	0.1	0.7	0.0	0.8
	Others	0.6	1.2	0.3	2.1
	White	0.1	4.0	0.4	4.5
Tshwane	Asian	0.0	1.9	0.5	2.4
	Black African	2.6	27.5	4.3	34.5
	Coloured	0.2	0.9	0.1	1.2
	Indian	0.0	0.7	0.2	0.9
	Others	0.1	2.0	0.3	2.4
	White	0.0	3.1	0.1	3.2
eThekwini	Asian	0.6	0.3	1.0	1.9
	Black African	9.4	4.0	7.9	21.4
	Coloured	0.1	0.0	0.0	0.1
	Indian	0.0	0.0	0.0	0.0
	Others	0.1	0.0	0.1	0.2
	White	0.0	0.2	0.0	0.2
Total		15.6	64.7	19.7	100.0

Source: Author (fieldwork, 2022).

higher proportion of black migrants (9.4%) feel restricted or limited in their movement in the eThekwini metropolitan area than in any other metropolitan area. In Tshwane and Cape Town, only 2.6% and 1%, respectively, of the black migrants rated the restriction or limitation on their movement as high. In this regard, figures for other racial groups of migrants are less than 1% and hence comparatively insignificant. Table 11.4 shows that with regard to

Table 11.4 Spatial Integration into Local Communities

	Spatial Integration of Migrants into Local Communities	High (%)	Low (%)	Moderate (%)	Total (%)
Metropolitan	Race				
Cape Town	Asian	1.4	0.9	0.8	3.1
	Black African	11.4	1.6	6.3	19.3
	Coloured	0.9	0.6	0.2	1.7
	Indian	0.6	0.1	0.1	0.8
	Others	1.0	0.5	0.6	2.1
	White	3.5	0.3	0.7	4.5
Tshwane	Asian	2.1	0.0	0.3	2.4
	Black African	21.6	2.2	10.7	34.5
	Coloured	0.9	0.0	0.3	1.2
	Indian	0.3	0.1	0.5	0.9
	Others	1.5	0.0	0.9	2.4
	White	2.6	0.0	0.6	3.2
eThekwini	Asian	0.0	1.9	0.0	1.9
	Black African	3.0	12.4	6.0	21.4
	Coloured	0.0	0.0	0.1	0.1
	Indian	0.0	0.0	0.0	0.0
	Others	0.1	0.0	0.1	0.2
	White	0.0	0.0	0.2	0.2
Total		50.9	20.7	28.5	100.0

Source: Author (fieldwork, 2022).

spatial integration, a significantly lower proportion of the black migrants (3%) in eThekwini rated the level of acceptance of foreign migrants by locals as high. By contrast, in Tshwane and Cape Town 21.6% and 11.4%, respectively, of the black migrant population rates the level of acceptance of foreign migrants into local communities as high in their metropolitan area. These statistics suggest that black migrants feel less welcome in eThekwini than they are in Tshwane and Cape Town.

11.3 Freedom of Access to Public Areas and Services

Figure 11.5 depicts migrants' responses to a number of questions that sought to establish the level of inclusion of migrants in South African cities with reference to access to public spaces. Amongst other things, migrants were asked the extent to which they commute public spaces without any fear of intimidation, hostility and xenophobia. They were also asked to rate a number of things such as, the level of freedom of access to public spaces as well their chances of being victimised as a foreigner in a public space.

As noted in Figure 11.5, 17% of migrants rated their level fear of hostilities in public spaces as low. By contrast, about 27% expressed a moderate level of fear while 55% expressed a high level of fear. Notwithstanding this, most migrants (about 67%) rated the likelihood of an attack in a public space as low or very low. This is congruent with the expressed level of freedom to

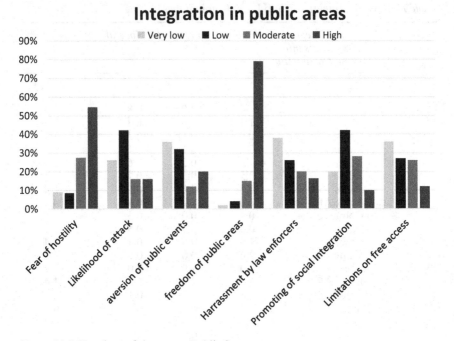

Figure 11.5 Freedom of Access to Public Spaces
Source: Author, (Fieldwork, 2022).

access public, almost 78% of the migrants indicated that the level of freedom to access public areas is high. While some of the migrants reported being harassed by law enforcers in public areas, only about 38% rated the level of harassment as moderate and high. On the other hand, about 62% of the migrants rated the level of harassment as low.

Although foreign migrants expressed some fear of hostility in accessing public spaces, overall they have considerable freedom to access such spaces. And in their estimation the likelihood of them being attacked in public spaces because they are foreigners is low. Some challenges to foreign migrants' access to public spaces include harassment by law enforcers. However, the harassment does not appear to be a major impediment for the inclusion of migrants in accessing public spaces.

11.4 Freedom to Access Service Centres and Amenities

Figures 11.6–11.8 allude to the reactions of migrants on access to service delivery.

About 15% of migrants indicated that the level of freedom and availability to use public transport is low. By comparison, about 83% of the migrants indicated that the level of freedom and availability to use public transport is moderate to very high. Three quarters of the migrants indicated that they do not fear being attacked while using public transport. This is highly correlated

Inclusion of Foreign Migrants in South African Cities   179

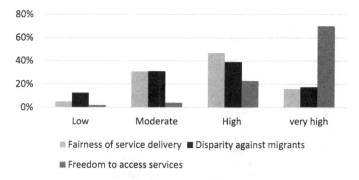

Figure 11.6 Fairness in Service Delivery
Source: Author, (Fieldwork, 2022).

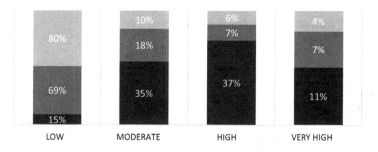

Figure 11.7 Freedom of Movement
Source: Author, (Fieldwork, 2022).

Figure 11.8 Word Cloud Propositions for Improving Spatial Integration

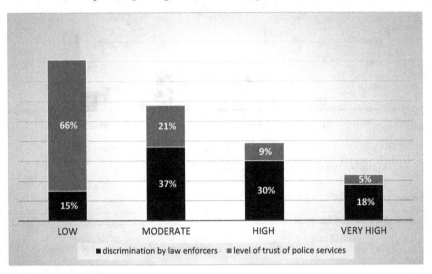

Figure 11.9 Trust of Law Enforcement Agents
Source: Author, (Fieldwork, 2022).

to the proportion of migrants who indicated that being attacked in public transport is not a common phenomenon (r = 0.92; P < 0.001). Hence, it can be deduced from these statistics that foreign migrants are included in South African cities through access to public transport. This is supported by observations on the ground where foreign migrants have access to, and they do use, public transport such as taxis and metro rail.

Figure 11.9 shows that there is trust deficit between foreign migrants and law enforcement agencies. About 66% of the migrants rated their trust in law enforcement as low. Only 35% of the migrants rated their trust of law enforcement agents as moderate to very high. Eighty-five per cent of the migrants indicated that discrimination by law enforcers is moderate to very high.

Table 11.5 shows the correlations between key variables that influence the choice to isolate, and job opportunities relative to migrant safety and security. Only the interactions that were significant at P < 0.01 are presented. Frequency of attacks, though significant, was weakly correlated to 'willingness to co-exist' (r = 0.26), and hostile behaviour of locals (r = 0.19) and the choice to isolate hence Tshwane and eThekwini show higher levels of co-existence in local communities. Frequency of attacks and xenophobia were hence significant and positively correlated (r = 0.77; P < 0.001) and reflect negatively on migrant opportunities for work (r = 0.82; P < 0.001). The disparities were more glaring with race, as Black Africans indicated that there were more prone to attacks (r = 0.98; P < 0.0001) and targets of xenophobia (r = −0.97; P < 0.0001). The age of migrants was not strongly related to most variables except for economic functions (job opportunities), which explained 84% of the variation. Overall, migrants seemed to enjoy freedom of movement regardless of the negativity. The 92% correlation between race and

Table 11.5 Correlation Matrix – Spatial Integration

	Willingness to Co-exist with Locals	Hostility by Locals	Choice to Isolate	Lack of Jobs	Xenophobia	Freq. of Race Attacks
Hostility by locals	(−0.58)	*				
choice to isolate	0.51	(−0.65)	*			
Lack of jobs	0.29	(−0.82)	0.42	*		
xenophobia	(−0.73)	0.79	(−0.61)	(−0.74)	*	
Freq. of attacks	0.26	0.19	0.23	0.39	0.77	*
Freedom of movement	0.65	0.14	0.02	0.25	(−0.21)	(−0.27) 0.03
Race	0.92	0.94	0.23	0.56	(−0.97)	(−0.98) *
Age	0.23	0.58	0.07	0.84	0.12	0.54 *
Spatial integration	0.27	0.36	(−0.92)	*	0.81	* 0.78

* – P < 0.01
Source: Author (fieldwork, 2022).

willingness to co-exist can be ascribed more to Black Africans that originate from Southern Africa. The exception is in Cape Town where Chinese nationals have set businesses in township and live with locals.

Table 11.6 shows the interrelationships between migrant status variables of race, gender, age, language and professional status and access to public resources and safety and security. Access to public services was not limited by race ($r = -0.13$) and correlations with gender, age and language were not significant (ns). However, access to services that require payments was strongly correlated with the professional status of the migrant, which can be explained by level of affluence, which gives an advantage to migrants that are in the country by choice and have competitive skills over political refugees. There were strong and inverse relationships between race and security and safety ($r = -0.59$; $P < 0.001$) which also had a strong link to 'lack of trust of security agents' ($r = -0.71$; 0.001) and the fear of discrimination ($r = -0.73$; $P < 0.001$). These significant inverse correlations with race are noted in Table 11.6, the driver being the Black Africans. Gender was significant, indicating that there were differences in levels of how Black females and males responded or were targets of security systems. With age, negative and significant correlations were noted in the variables 'discrimination' and 'risk of victimisation'. The correlations were inverse and highly significant ($r = -0.66$; $R = -0.83$; $P < 0.0001$). The most striking correlations were related to language, which had strong but inverse relations with security and safety variables. For black Africans being able to speak the local languages including Afrikaans is an enabler for access to public resources and integration.

182 *Inclusion of Foreign Migrants in South African Cities*

Table 11.6 Correlation Matrix – Security and Safety

	Race	Gender	Age	Professional Status of Migrants	Language of Migrants
Access to public spaces	(−0.13) ns	0.06 Ns	0.071 ns	0.63 0.001	0.12 ns
Access to community services	0.53 0.001	0.1 Ns	0.19 ns	0.77 0.001	0.23 0.05
Access to security & safety	(−0.59) 0.001	0.22 0.01	0.18 ns	0.82 0.001	(−0.83) 0.001
Risk of victimisation	(−0.62) 0.001	(−0.49) 0.01	(−0.83) 0.0001	(−0.39) 0.001	(−0.89) 0.001
Lack of trust of security	(−0.71) 0.001	(−0.53) 0.01	0.33 0.05	(−0.58) 0.001	(−0.92) 0.0001
Random checks by security	(−0.69) 0.001	(−0.64) 0.001	(−0.59) 0.01	(−0.65) 0.001	(−0.43) 0.005
Fear of discrimination	(−0.73) 0.001	0.48 0.001	(−0.66) 0.0001	(−0.78) 0.0001	(−0.86) 0.0001

Source: Author (fieldwork, 2022).

Figure 11.10 Security Concerns of Foreign Migrants

11.5 Inclusiveness through Access to Public Security and Safety

Figure 11.10 shows a word cloud that highlights some of the sentiments expressed by migrants when asked to rate their level of comfort in approaching the police on any issue of safety and security. Words with negative connotations such as 'corruption', 'corrupt officers', 'bribes', 'abuse of power' and 'biased' feature prominently and frequently in the word cloud. There are also some positive words such as 'helpful', 'approachable', 'professional' and 'satisfactory'. Hence, it seems some migrants have a negative perception of the police. They believe there are some bad elements within the police force that are corrupt and/or who abuse their power when dealing with foreign migrants. Indeed, there have been press reports of police demanding or accepting bribes from illegal immigrants or, targeting shops belonging to foreign migrants. Such incidences have the effect of making migrants feel excluded because public security and safety officials appear to work them. However, not all foreign migrants share this negative perception of the police since, as noted earlier, there are some positive words in the word cloud. These positive words suggest that some migrants believe that police are doing a good job. Such migrants would most likely feel included through access to public safety and security as they perceive the police to be safeguarding their safety and security.

11.6 Inclusiveness through Access to Economic Opportunities

'Employment is a core part of the integration process and finding a job is crucial to becoming part of the host country's economic and social life and thus developing a sense of belonging to the host society' (EESC Study Group on Immigration and Integration, 2020). To gauge the inclusiveness of South African cities in relation to foreign migrants' access to economic opportunities, a number of questions with a bearing on access to employment opportunities were posed to migrants. At least 52% of the migrants indicated that they had found employment in the informal economy and, a majority of these are self-employed. About 48% found employment in the formal economy. Several factors could explain why most migrants end up in the informal economy. A foreign migrant cannot legally work in South Africa unless they have the relevant work permit or permanent residence permit. Given that, as observed elsewhere in this study, the DHA takes a long time to process permits and other documents, most immigrants find themselves having to engage in informal employment – often illegally – just to earn an income to survive. It could also be that due to lack of recognition of some foreign qualifications, some migrants find themselves unable to secure jobs in fields they trained for; hence, they have to make do with whatever opportunities they come across.

About 36% of the migrants are employed in the government. This is a significant figure, and it speaks to inclusivity. Figures 11.11 a, b and c show that 28% of migrants are unemployed. According to the word cloud (Figure 11.12), it appears the positions occupied by migrants cover a broad spectrum of industries and professions. These include 'mechanical engineers', 'accountants', 'nurses', 'teachers', 'researchers', 'professors', 'lecturers' and 'chefs', just to name a few. Those that are self-employed are active in diverse sectors. There are 'general traders', 'clothing retailers', 'auto mechanics', 'tutors', 'hairdressers' and 'delivery'.

184 *Inclusion of Foreign Migrants in South African Cities*

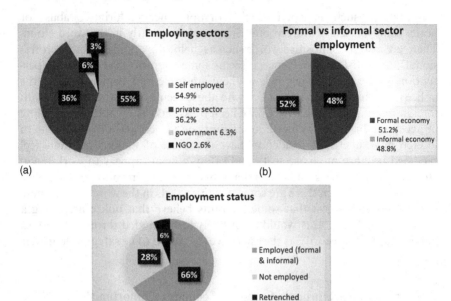

Figure 11.11 (a) Employment Sector, (b) Employing Institutions, (c) Employment Status

Source: Author, (Fieldwork, 2022).

Figure 11.12 Employment Challenges

Failure to secure employment was linked to non-recognition of foreign qualifications, about 81% rated the possibility as moderate to high. Migrants must submit their foreign qualifications to SAQA for verification and evaluation. SAQA will then issue them with a certificate to certify which South African qualification the foreign qualification is equivalent to. Most employers accept the certificate issued by SAQA as proof of one's qualifications. It may be that some migrants are not aware of SAQA, its role and processes in certifying foreign qualifications or prejudice by institutions. In which case there is need for civic organisations/NGOs to create awareness among migrants about SAQA and to advise migrants of the process to get their qualifications verified and evaluated (Figures 11.13 and 11.14).

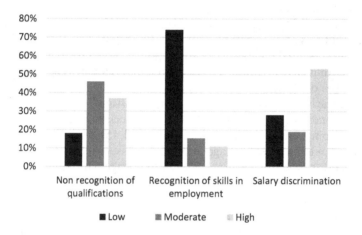

Figure 11.13 Economic Integration

Source: Author, (Fieldwork, 2022).

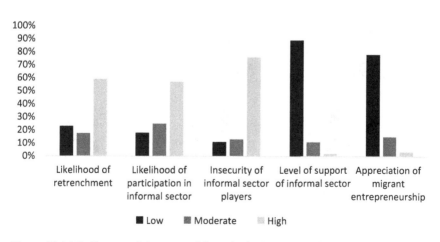

Figure 11.14 Indicators of Access and Security in the Economic Sector

Source: Author, (Fieldwork, 2022).

11.7 Institutional Support and Immigration Policies

Civic organisations/NGOs play an important role in integrating migrants in their new environments. They offer support and guide migrants through the integration processes (EESC Study Group on Immigration and Integration, 2020). In this study, migrants were asked whether they were aware of any institutions/NGOs that assist migrants and, if they knew any such, whether they were satisfied with the services provided by the institutions. They were also asked to rate whether such institutions had any impact in influencing the development of immigration policies. Fifty-one per cent of the migrants rated their level of awareness of civic organisations/NGOs that assist migrants as low. About 28% rated their level of awareness as moderate, while 23% rated it as high. Hence, it would appear that there is a fair level of awareness of the existence of such organisations. However, the level of satisfaction with the interventions of civic organisations/NGOs is quite low (72%). Only 28% of the migrants rated their level of satisfaction as moderate to high. In rating the impact that civic organisations have on the development of immigration policies, 70% of the migrants rated it as low while 30% rated it as moderate to high. Clearly, there is room for civic organisations/NGOs to do more to create awareness among migrants about the services that they offer and their role in shaping immigration policy (Figure 11.15). It is important that migrants should have centre stage in projecting issues affecting them; hence, it is crucial for civic organisations/NGOS to engage with migrants in order to effectively assist them to represent themselves (EESC Study Group on Immigration and Integration, 2020).

Figure 11.16 captures the ratings assigned by migrants to various aspects of their interphase with the Department of Home Affairs. For example, in one question migrants had to rate the services provided by the department

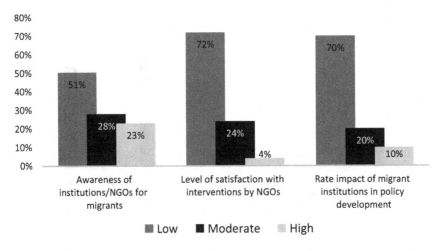

Figure 11.15 Awareness of Institutions Supporting Migrants

Source: Author, (Fieldwork, 2022).

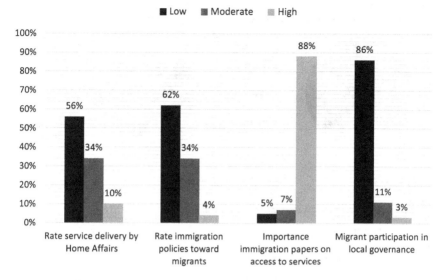

Figure 11.16 Home Affairs and Local Governance
Source: Author, (Fieldwork, 2022).

of homes affairs to migrants. In another, they had to rate South Africa's immigration policies towards migrants in general. They also had to rate the importance of having immigration documents to access services. About 56% of the migrants rated the service delivery from Home Affairs as low. They were concerned that the department does not offer a quick service. About 34% rated the service delivery from Home Affairs as moderate while 10% rated the service as high. Concerning the importance of immigration documents such as visas in accessing services, about 88% of the migrants rated the importance of such documents as a high. In other words, possession of such documents is critical for accessing services. Only 12% of the migrants did not consider such documents essential for accessing services. In light of the foregoing, it is clear that poor service delivery by Home Affairs negatively impacts inclusivity. There is need for Home Affairs to review its service delivery to see how they can improve so that foreign migrants that are legally entitled to have access to certain services are not prejudiced because the department is delaying or failing to issue them with the relevant documents.

There is a significant inverse correlation between immigration policies, service delivery and access to basic services ($r = 0.57$; $r = 0.61$; respectively $P < 0.01$) as 62% rated that immigration policies were not favourable and visa or related documents issued by Home Affairs were a hindrance to accessing basic services. Consequently, this limits participation of foreigners in local governance as noted by 86% of respondents. Documents issued by Home Affairs are considered critical (88%).

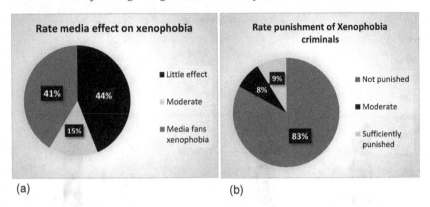

Figure 11.17 Foreigners' Views on Xenophobia

Source: Author, (Fieldwork, 2022).

11.7.1 Role of Media in Xenophobia

Migrants were invited to rate if they strongly believed the media has contributed to escalating the public's wrong perception about migrants. The ratings appear in Figures 11.17. Forty-four per cent of the migrants felt that the media had little effect on the public's perception of migrants. Fifteen per cent said the media had a moderate effect on while 41% thought the media escalated public's wrong perception of foreigners. Thus, the general perception among 56% of migrants is that the media's contribution in escalating the public's wrong perception about migrants is high. With regard to punishment of perpetrators of xenophobic comments or attacks, an overwhelming majority of 83% thought that the perpetrators are not punished while 9% were of the view that the perpetrators are not sufficiently punished. Only 8% of the migrants thought the perpetrators were moderately punished. These statistics are of concern. A well-known adage demands that justice must not only be done but must also be seen to be done. In the eyes of the migrants, justice does not seem to be done when it comes to punishment of perpetrators of xenophobic comments and attacks. Such a scenario does not engender feelings of being included.

11.8 Inclusivity and Integration

Migrants were asked to rate the level of inclusivity in their metropolitan area in relation to different aspects such as spatial integration, access to public spaces, access to public services and supplies, education, etc. Access to public spaces and spatial integration were the two areas migrants rated as having high level of integration. This conforms to what is observable on the ground. Even though some migrants may be apprehensive about living among the locals, there is evidence that migrants have integrated very well into local communities and they co-exist with locals in the townships. Moreover, locals are able to access public spaces such as parks, libraries and shopping malls without inhibition. Almost 97% of the immigrants indicated that the lowest

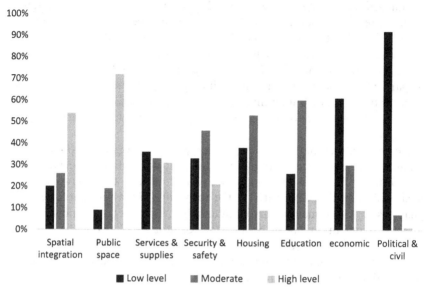

Figure 11.18 Inclusive across Sectors

Source: Author, (Fieldwork, 2022).

level of inclusion is in the area of political and civil participation. This is understandable as the whole world over political and civil participation in activities such as voting in general elections as generally reserved for a country's citizens. Inclusivity through access to economic opportunities was also rated as very low by about 57% of the migrants. Only about 33% of the migrants rated inclusivity in that sphere as moderate to high (Figure 11.18).

11.9 Conclusion and Recommendations

The objective of this study was to establish the extent to which foreign migrants have been physically, economically, socially and politically integrated in South African cities. It is clear from the research findings that there is a moderate to high level of inclusiveness in several areas, including spatial integration, access to public spaces, education, housing as well as safety and security. This high level of integration is likely a result of what appears to be South Africa's migration policy that can be regarded as one of social inclusion and integration. In this regard, South Africa appears to have a policy of allowing migrants, including asylum seekers, to integrate into local communities. Consequently, migrants especially those from the SADC have largely settled among the locals in the townships with some even intermarrying with the locals. This policy of social inclusion and integration contrasts sharply with the policies followed in neighbouring countries like Zimbabwe and Malawi. Zimbabwe follows the refuge encampment policy whereby asylum seekers are required to stay in refugee camps and not integrate into the local communities (Mhlanga and Zengeya, 2016). Malawi also appears to follow a similar refugee containment policy whereby refugees are required to reside in the designated refugee camp, Dzaleka Refugee Camp (Mvula, 2009). Hence, South Africa's policy permits migrants to integrate into local communities. It allows them freedom of movement and gives them opportunity to participate in the economy through employment and undertaking business ventures. Indeed, migrants from countries such as Somalia, Pakistan and Bangladesh are known to have set up businesses such as spaza shops in the townships.

Despite the fact that foreign migrants are free to integrate into local communities, some sections of the foreign migrant population, especially those from outside the SADC region, have found it difficult to integrate into the local communities particularly in the townships. Such migrants cited language as an impediment to spatial integration. They feel uncomfortable and insecure living among locals because they cannot speak any of the local languages. Hence, they prefer to live among their fellow migrants instead of integrating into the local communities. As a result, they also feel limited or restricted in their movement within cities because of their inability to fit in to the local communities due to language issues.

> **Recommendation 1:** It is recommended that the government, working together with non-governmental organisations (NGOs) that assist migrants, must develop a structured language training program to enable migrants to receive training in one or more local languages.
>
> **Recommendation 2:** The government, in collaboration with NGOs, must craft a holistic national integration strategy, including the development structured services – such as the language training program in Recommendation 1 – aimed at facilitating social integration. These structured services should form part of standardised integration program.

Even though migrants are free to integrate into the local communities, some of the migrants reported that there is very little government support to assist them with the process of settling in. Migrants are left to their own devices to find a way to become economically viable. This creates challenges for some of the migrants to sustain themselves economically. This is especially so for the undocumented migrants because legally, they are not supposed to work or engage in South Africa unless they have a valid work permit or other relevant permit. Further, most of migrants are not entitled to state sponsored social support programs such as monetary grants like the child support grants, disability grants, dependency grants, foster care grants, etc. Whatever support migrants receive largely comes from non-governmental organisations (NGOs) such as churches and other donor organisations. However, the research results showed that not many migrants were aware of the NGOs that assist migrants. Those that were aware of such organisations had reservations concerning the effectiveness of such organisations in assisting migrants.

Recommendation 3: The government and NGOs must work together to create awareness about the existence of NGOS that assist migrants with the process of integration.

Recommendation 4: Since NGOs are at the coal face working with migrants, they are well acquainted with the issues affecting migrants concerning integration into South African society. As such, it is recommended that government must work closely with these organisations to come up with migration policies that address issues faced by migrants.

Another area of concern where there is room for improvement in terms of migrant inclusion is the area of safety and security. Although the level of integration with regard to safety and security could be considered moderately high, there appears to be a trust deficit between migrants and the police department. This trust deficit is borne out of the perceived corrupt activities and unprofessional conduct of some police officers towards foreign migrants. As a consequence, some foreign migrants feel that they cannot count on the police to ensure their safety and security; this impacts negatively on the integration of migrants into South African society.

Recommendation 5: It is recommended that police must investigate cases of xenophobia thoroughly without fear or favour and ensure perpetrators of xenophobia are brought to account.

Recommendation 6: It is recommended that police undertake internal investigations on issues of corrupt officers who demand bribes from foreign migrants.

Dissatisfaction with service delivery by the department of DHA was one of the issues that stood out from the feedback given by migrants. They complained that the department takes long to process applications and issue permits. This creates challenges for migrants, especially asylum seekers, because they have to wait for long periods of time before the department resolves their applications for asylum. During the waiting period migrants sit idly at home because legally they are not entitled to work in SA. This creates despondency and makes integration difficult.

Recommendation 7: Department of Home Affairs must commit and adhere to a specified and reasonable turnaround time for processing relevant migrant permits. This will ensure migrants have a clear indication of when they can expect their applications to be dealt with.

Recent history as well as results from the study shows that anti-immigrant sentiments and xenophobic attacks against migrants are a reality in South African society. These anti-immigrant sentiments manifest in xenophobic attacks against foreigners and in movements such as Operation Dudula. Common complaints against foreigners are that they are taking away economic opportunities from locals and, that there are too many illegal migrants who commit crime in the local communities they settle in (Aljazeera, 2022). Unfortunately, these complaints are not based on any scientific evidence; they are just generalised statements presented as facts. No doubt, there are migrants that are in South Africa illegally and some of whom have been involved in the commission of crime. However, it is highly debatable whether the presence of migrants – legal or illegal – is the dominant causal factor of the socio-economic problems faced by South Africa such as high-income inequality, high rate of unemployment, poor service delivery and high rate of crime. Scapegoating migrants as the cause of South Africa's social ills heightens anti-immigrant sentiments and may lead to more xenophobic attacks on foreigners when what is needed is empathy for migrants.

There is need to create awareness among locals of the reasons why migrants leave the familiar environments of their home countries coming to South Africa and the plight they face as they try to take care of themselves as well as their families back home. The government must redouble its efforts to address the socio-economic issues faced by locals without neglecting or demonising migrants.

By and large, there is integration of foreign migrants across most variables considered for this study with the exception of political and economic sectors. While from the migrant's perspectives there is an expectation that the government should do more in assisting with integration, it must be understood that the South African government cannot cater for each and every persons' need. The government already has challenges providing for its citizens. Perhaps there is a greater need for NGOs to play the intermediary role to provide support and assist with migrant integration.

Bibliography

EESC Study Group on Immigration and Integration. 2020. Project on the role of civil society organisations in ensuring the integration of migrants and refugees. *European Economic and Social Committee*, Greece, 5–6 March 2020. Available online: https://www.eesc.europa.eu/en/our-work/publications-other-work/publications/project-role-civil-society-organisations-ensuring-integration-migrants-and-refugees-2

Landau, L. B., Segatti, A. & Misago, J. P. 2011. *Governing Migration & Urbanisation in South African Municipalities: Developing Approaches to Counter Poverty and Social Fragmentation*. South Africa Local Government Association (Salga).

Mhlanga, J. & Zengeya, R. M. 2016. Social Work With Refugees in Zimbabwe. *African Journal of Social Work*, 6, 22–29.

Mvula, L. D. 2009. Refugee Status Determination and Rights in Malawi. Refugee Studies Centre Workshop Discussion on RSD and Rights in Southern and East Africa, Uganda.

Myeni, T. 2022. *South African anti-immigration vigilante leader granted bail*. Aljazeera, 28 March. Available online: https://www.aljazeera.com/news/2022/3/28/south-african-anti-immigration-militia-leader-granted-bail (Accessed 21/12/2022)

12 Reflections on Immigration Policies and Inclusivity in Southern African Cities

12.1 Introduction

This book provided an in-depth overview of the selected major cities of Southern African with the intention of understanding the status of immigrants. It is built on the premise that immigrants are among the vulnerable members of society who are always perceived as outcasts in cities. While this perception of exclusivity prevails in no uncertain terms, its depth and levels of prevalence varies among the cities discussed in this book. However, what is clearly evident is that the notion of immigrant inclusivity in these cities is a 'vicious contested space' which pits diverse stakeholders (such as civil society, governments, immigrants and host communities). The authors took an institutional approach in analysing the institutional frameworks of these cities in relation to the governance of immigrants. The overall intention was to establish the extent to which the institutional framework is accommodative of these immigrants. Emerging evidence discussed in various chapters shows various trends and approaches in the administration of immigrants which in turn weigh differently on their inclusivity in urban areas.

12.2 Diversity of Institutions

It was indicated in the introductory chapter that in as much as cities of Southern Africa are united by similar historical circumstances which moulded them, there are still finer nitty-gritties which usher diversity in their institutional approaches in relation to immigrants. The search for individual identity evolved over years and at each stage in the evolution process, institutional frameworks were refined appropriately. Therefore, what we see today in these cities is a continual refinement of immigration laws in response to contemporary social-economic pressure and geo-political dynamics. Key strands emanating from the country profiles in the various chapters seem to zero in on the following factors.

12.2.1 Migration Laws

An overview of all cities discussed in this book shows that there is a strong regime of policies, laws and regulations that prevail in relation to migration. These are shaped by both international and regional protocols (such as the Refugee Convention of 1951, the Protocol Status of Refugees of 1976 and the Organisation of African Union of 1969, among others), and contextually tailor-made in line with each country's socio-economic and political disposition. Among other laws that prevail in each context, the Constitution serves as the highest law of each country and underlines the sanctity of human life (irrespective of background, race or nationality) through upholding of human rights, dignity, equality and freedom. This forms the entry point of migration laws at the national level but administrative mechanisms put in place to manage immigrants with diverse intervention mechanisms ranging from an '*open-door-policy*' to a policy of '*strict control*'. It must be understood that the sovereignty of any country is determined by law and order – therefore, laws of entry and exit in each sovereign state are critical to safe-guarding itself. Sovereignty, as Newton and Van Deth (2016) perceive it, is about having control over territorial space, which in turn provides control over resources and legitimate use of enforcement mechanisms (among others). The operation of these policy mechanisms through various regimes of regulations is largely responsive to the status of immigrants, as elaborated below:

Legal immigrants: What is clear from all cities is the fact that all legal/documented immigrants who qualify for entry (whether on permanent or temporary basis) are protected by the host country's constitutional obligations. As such, the extent of these immigrant rights differs from country to country and is mainly responsive to the status of immigrants (such as holders of permanent resident permits). The level of benefits such as access to employment and educational facilities are commensurate with the conditions that are spelt out on the permit.

However, this is in direct contrast to another dimension of legal migrants who come in the form of refugees. With the exception of South Africa and Mozambique, which seem to provide an open-door policy to immigrants of diverse backgrounds, other countries seem to have strict control mechanisms which impact negatively on people who fall within the category of refugees. Zimbabwe, Malawi and Botswana's immigrant policies restrict refugees to spatially designated areas known as refugee camps. It should be understood that in the terminology of foreign/international policy, refugees are perceived as stateless people whose tenure in the host country borders on seeking asylum. Until such a stage that they qualify as immigrants under any prevailing regimes of the host country's stabilising policy for immigrants, their status will always remain as temporary residents '*in transition*'. Hence, they are a responsibility not only of the host country but also of the international community such as the United Nations High Commission for Refugees (UNHCR).

Illegal migrants: Despite border control measures and other enforcement regulations in place, cities of Southern Africa continue to experience influxes of undocumented immigrants. Such migrants, as will be elaborated below, are a result of socio-economic and political dynamics. All cities are experiencing this problem, and it is more pronounced in countries where an open-door policy prevails (such as in South Africa and Mozambique) but more so, where economic opportunities are promising (such as in South Africa and Botswana). In the case of Johannesburg in South Africa, the lobbyist and dialogue approach which is used to deal with such migrants has aggravated the situation as more people continue to flock to the city and the country at large amid deportations. Other countries such as Botswana (and to some extent Zimbabwe), employ a policy of restriction which includes mechanisms such '*arrest and* deport' as well as '*stop and search*' in a bid to clamp down on illegal migrants. This have rendered immigrants to '*cry victim*' irrespective of the fact that they are the ones who flout host countries' immigration laws.

But what clearly emerges from the narrative of city profiles vis-a-vis immigrants is that this is becoming a major problem since illegal migrants are on the increase. No country is prepared to support 'undocumented' people who continue to impact negatively on national resources and infringe on the country's sovereignty.

12.3 Geo-politics and Immigrants

What is also emerging is that migration and institutional frameworks that are at work are dynamic. Cities of Southern Africa have been undergoing significant changes in response to the geo-politics of the region, Africa and beyond. These have been evolving throughout the colonial and post-colonial period and have impact on the status of immigrants.

The traditional open-door policy (alluded to above) was a common phenomenon during the colonial period when the search for labourers and skilled manpower was on demand in some of these cities. The labour train, which connected Johannesburg to Lusaka via Gaborone and Bulawayo (in Zimbabwe), was symbolically evident of the open-door immigration policy which allowed labourers to cross borders to work. This was a selfish mechanism put in place by colonial regime meant to benefit global imperial projects such as the Federation of Southern Rhodesia and Nyasaland and the goldfields of South Africa. The formal stabilisation of immigrant labourers was left hanging and did not seem to be the ultimate goal as witnessed in Zimbabwe's Fast Track Land Reform Programme (of 2000) when farm labourers from Malawi and Mozambique came to witness.

On the other hand, skilled manpower in the form of engineers, doctors and teachers seem to have a fair deal when it comes to migration as they go through legal channels of immigration laws. The colonial and post-colonial periods have shown that most of these professional migrants easily assimilate in societies of host countries. Their educational status and ability to contribute to the host country's socio-economic status has seen some countries such as South Africa providing special immigration dispensations to Zimbabwean with relevant skills. Similarly, in Lusaka, Zambia, the Chinese, who come with '*investment clout*' experience better reception.

Southern African cities are also recipients of people experiencing forced-migration in the form of refugees. Historically, ravaging political instabilities saw countries opening their doors to migrants fleeing the ravage of instability in Zimbabwe, Mozambique and South Africa. This, unfortunately, is continuing as refugees are coming from beyond the borders of Southern African cities (such as the Horn of Africa and Central Africa). The situation has been aggravated by economic downswings in countries like Zimbabwe (since the inception of the Fast Track land Reform Programme in 2000 – see Chipungu and Magidimisha (2021); Chipungu and Adebayo (2012)), prompting the up-surge of both migration into neighbouring countries and beyond. It is under such circumstances that immigration laws are shifting in response to these dynamics. Hence '*protective immigration policies*' (such as those prevailing in Zambia and Botswana) are becoming more pronounced as countries try to clamp down on illegal migrants, minimise their expenditure on migrants and, overall, safe-guard their national interests. In cities such Johannesburg, societal conflicts in the form of xenophobic clashes are pointers to the level of attrition happening at the city level between immigrants and the host community.

Lastly, the economic status of the host country and its cities play a critical role in migration for they influence the motives of migrants. Countries such as Malawi and Mozambique are not perceived as final destinations by some migrants especially those emanating from other African countries. Most migrants are seeking economic freedom even when they are fleeing from political instability. Hence for that reason, they see cities such as Lusaka, Lilongwe and Maputo as transitional zones in their bid to economic freedom specifically to South Africa and abroad. It does not matter how many years they spend in the refugee camps of these countries because for them, they will be mobilising resources to move on. South Africa's '*open door policy*' coupled with access to economic and social services irrespective of immigration status provides the most lucrative final destination.

12.4 Immigrants and Inclusivity in Cities of Southern Africa

The city has two major dimensions – the physical dimension and the human dimension. The two dimensions are in a complex and dynamic relationship which epitomises the city as a living organism. Hence, the city, like any other organism, requires a state of balance in order to achieve sustainable existence now and tomorrow. West (2018) further sheds light on this argument by alluding that cities, like any other evolving systems need continual supply of energy. A critical factor which affects the city is urbanisation which manifests itself through demographic and spatial dimensions. In the same line of resonance, immigration is a critical input into the city's spatial and demographic systems which have to be monitored throughout the city's life. The only way to sustainably balance this component of immigration is through inclusivity, which if not rightfully executed has negative repercussions. These issues were briefly discussed in Chapter 1 and Chapter 3 and are revisited in this chapter.

The point of entry into this discussion is to track pathways of exclusion emanating from the case studies of cities in Southern Africa in line with the conceptual framework provided in Chapter 2.

12.4.1 Immigrants and the Economic Platform

As already noted in Chapter 2 under international economic systems, among other factors that trigger migration are economic factors. Similarly, the stabilisation of immigrants in host countries is governed by both economic labour laws among other factors requirements. What emerges from the various case studies is that inclusivity of immigrants into the economic fold of the host country is largely driven by the status of migrants coupled with the prevailing labour-economic laws at work. In addition, this also depends on each country's economic status.

From a purely historical perspective (as already noted above in the preceding section under geo-politics), the need to import labour to quench the industrial, mining and farming needs of host countries saw cities such as Harare and Johannesburg using liberal migration policies that enabled them to capitalise on labour requirements. Both general and skilled workers were easily assimilated into the host society. The post-colonial stage in most countries still promotes immigration of skilled labour which contributes to the growing economies of these cities and their respective countries. These policies have continued even in post-colonial times but with more focus on skilled manpower as opposed to general labourers. However, even for skilled manpower, immigrants have to endure strict screening processes in order to qualify for visa requirements. Special working dispensations, like in the case of Zimbabweans in South Africa, is clear evidence of how inclusive some of these labour immigration policies are as they are meant to ensure economic growth and continuity.

Another emerging economic dimension that reflects the issue of inclusivity is observable in the case of Lusaka, where the need for economic investment has seen an increase in Chinese immigrants. Such economic immigrant policies are more receptive to investors since they contribute to the growth of the economy of the city while at the same time offering employment opportunities for local people. Similar trends are also observable in Lilongwe, Maputo and Johannesburg, where a sizeable number of Asians are also economically active.

However, this is in direct contrast to refugees whose 'fluid status' makes it difficult for many countries to confer on them similar inclusive benefits. As already noted, in all cities (with the exception of Johannesburg and Maputo), immigrants are perceived as 'people in transit', and therefore, they are confined to refugee camps without access to economic benefits such as employment. This policy of restriction marginalises refugees to the extent that they are excluded from the job market. On the other hand, South Africa's 'open immigration policy' allows refugees to compete for economic opportunities at the same level with local people. However, it is only those awaiting for the outcome of their asylum application who are denied such opportunities. In most cases, these take up menial jobs such as in the construction and hospitality industries. Some refugees from West Africa, Horn of Africa and Asian countries have taken up downtown investment opportunities running their small businesses (such as hair salons and shops). Similar developments are observable in Maputo.

The last group of immigrants are those that are not documented. These they try by all means to evade any regulations, thereby ending in an informal economic environment where they are not recognised. At this level, they are highly marginalised and are excluded from various economic benefits. Harare, Gaborone and, to some extent, Lilongwe are very clear on this issue – there is no work for illegal migrants and this is enforced through the '*stop-search-arrest-deport*' policy. Gaborone takes a step further by raiding construction sites and premises where undocumented migrants work. Unfortunately, some of these people (in the case of Lilongwe and Lusaka) are '*refugees in waiting*' who are seeking means for survival. For those who work illegally, like in the case of Johannesburg, they are constantly exposed to extortion, underpayment and ill-treatment.

In summary, it can be emphasised that lack of economic inclusivity for immigrants undermines their livelihood status and also affects the city's economic integrity and performance. Immigrants are caught up in a web of economic informality where they engage in illegal activities, do not contribute financially to the city's economy and contribute to social ills and conflicts (as seen in the case of illegal miners in Johannesburg).

12.4.2 Immigrants and the Spatial Dimension

The spatial dimension to inclusivity in the city is achieved through access to land, housing and infrastructure. Immigrants are part of the human dimension of cities whose existence equally depends on access to land, housing and infrastructure. Access to these facets of the city is regulated. From a purely urban design perspective, the size of the urban population should be commensurate with the level of infrastructure, availability of housing and land to support current and future development. This explains why immigrants should be documented if sustainable development is to be achieved. More so, it also explains why certain quotas are placed on immigrants in line with the city and country's plans developments plan. For most documented immigrants, access to the spatial dimension is not a major challenge given that they enjoy benefits of the host city through access to employment which enables them to decent housing and infrastructure housing. Hence their status as documented and economically emancipated immigrants gives them the power of choice in terms of the quality of housing and infrastructure in terms of affordability.

On the other hand, refugees and undocumented migrants do not enjoy the same benefits. Evidence from Lusaka, Harare and Lilongwe show that most refugees are relegated to refugee camps where conditions are poor (such as overcrowding). Some of them have since left these refugee centres and got absorbed into informal environments where they lack recognition, houses are poor, lack physical infrastructure (such as water, electricity and sewer reticulation) and poor public transport. It is in these similar environments where most undocumented immigrants are found. The informal environment is an environment of obscurity which, in most cases, is not formally recognised on the city management plan. For this reason, it represents exclusion at its highest level. However, this unfortunately, is the norm in most of these cities, especially in Lilongwe and Maputo where cities are struggling to provide decent housing and supporting infrastructure.

However, in the case of Johannesburg and Gaborone, there are two points worth noting about undocumented migrants and those with refugee status. The high levels of exclusion force them into downtown where they stay in crowded environments sharing one or two rooms. This is a strategy to evade high rentals from property owners who overcharge them. The inner city is closer to opportunities and allows some of them to go into informal jobs such as trading in informal markets and other small business opportunities. One can therefore conclude that spatial exclusion is not only common to undocumented immigrants – but is a pandemic which host societies also experience as government struggle to provide decent services in the form of housing and infrastructure. All case studies in this context experience the same problem when it comes to the spatial dimension of inclusivity. The search for spatial equality among immigrants, especially those in the lower spectrum of the economy, is not a lone struggle which pits them against the host society. More so, it is not a pre-designed exclusionary policy that negates the rights of immigrants irrespective of their status – rather, it is a 'perpetual pandemic' whose impact does not discriminate between immigrants and the host society which governments are failing to tame.

12.4.3 Immigrants and the Social Dimension

The social dimension of inclusion rests on pillars of community participation, safety and governance. This is a dimension that hinges on societal recognition and government intervention through various structures. Therefore, the underlying factor is recognition of societal groups through formal and informal channels to enhance communication. This in turn impacts on immigrants from two perspectives – one being that immigrants are a minority group whose voice at times is muffled and lost in the midst of the host communities. It speaks to the levels of integration and assimilation of immigrants within society. Among documented immigrants, their levels of integration allow them to participate with the local community without any problems. In this regard, their economic status elevates them from vulnerability – an exclusionary label which is attached to most migrants. Their high levels of assimilation have made them part of the local community, thereby enabling participation at all forums.

On the other hand, undocumented immigrants and some refugees are among the vulnerable members of society whose status makes them 'social misfits'. For that reason, they are bound with fear that robs them of the freedom to choose housing environments, participate in community activities and move freely in search of opportunities. For them, *'choice'*, if ever it exists, is bound with *'conditionalities'* of survival both economically and physically as human beings. In other words, they have to weigh the situation by the *'balancing act'* of risking their life and achieving economic freedom. It is the need for survival that gives them the instinct to stay in areas where there are other foreign nationals in order to ensure safety as a group. Street battles that rocked Johannesburg in the form of xenophobic attacks are a pointer to the poor levels of cohesion and constant friction in host–immigrants relationship. Those immigrants who reside among locals still experience uneasiness emanating from cultural barriers (such as language barrier) which easily makes them social misfits. It is therefore not surprising that when tension erupts in such communities, immigrants become easy targets. Hence *'protection fees'* which come in the form of high rentals (to landlords) or occasional payments to locals serve to protect them from marauding locals who would take any opportunity to intimidate them. Above all, the continual search for economic survival pushes immigrants to the edge where some engage in illegal activities such as drug and human trafficking. These under-world activities, unfortunately, have tarnished the image of most immigrants who have all been *'inscribed into the web of criminals'* irrespective of their genuine survival strategies and social standing. The case of the *zama-zama* (illegal gold miners) in Johannesburg and spates of robberies in Gaborone are all pointers to the gravity of the negative activities of some immigrants.

Therefore, the lack of official identity, coupled with fear of harassment and deportation, has relegated undocumented foreign nationals to *'invisibility'* on the social platform. This social seclusion hampers their effort to participate and engage in various activities. Immigrants have come to bear the brunt of being labelled criminals and unwarranted competitors in cities where economic opportunities are continuously shrinking.

12.4.4 Immigrants and Levels of Inclusion

What emerges from this discussion is the observation that inclusivity among immigrants in cities of Southern Africa is hierarchical. This hierarchy is a function of a multitude of factors, among which are economic freedom, institutional frameworks and societal dynamics. It should be noted that this hierarchy is not cast in concrete. The levels of inclusion are very fluid and they respond to various dynamics at work. Therefore, within these cities, inclusion ranges from *'Level I to Level III'* with Level III being the highest level of inclusion.

12.4.4.1 Immigrant Inclusion Level I

Level I forms the basic level of entry into city inclusion by immigrants. It is the elementary level which is associated with undocumented foreign nationals. Because they suffer from lack of identity largely emanating from proper documents expected in the host country, they live a life in isolation mainly in the informal world characterised by informal connectivity to spatial, economic and social platforms. Unlike host citizens who might be languishing in the same fate, their situation is worsened by a lack of official recognition which in turn denies them formal employment, access to services and participation in community. As documented in the case of Gaborone, destitution, lack of basic necessities, lack of safety nets and ridicule from officials characterise the status of these immigrants. For them, this is a complete level of isolation whose minimal inclusivity depends on the informal channels of the host society and government organisation such access to health facilities. In real terms, this is a world of exclusivity experienced by undocumented immigrants and some refugees who languish in sheer marginalisation. However, this does not mean that within this same level, there are some undocumented immigrants who do not enjoy relative levels of inclusion in the host society.

12.4.4.2 Immigrant Inclusion Level II

At this level, formal recognition of immigrants opens doors for opportunities and participation on various platforms. Official recognition by cities and government officials translates to acceptance and participation in community. Access to housing in the form of refugee camps and urban residence cards (in Zambia) and access to educational facilities and employment in South Africa elevates the status of immigrants in host countries. To some extent, migrants have the freedom of choice to opportunities, movement and community participation. However, the status of refugees revolves around certain conditionalities which cap their progression socially and economically. For instance, most people with refugee status (in South Africa) cannot compete on the same level with locals for job opportunities – a situation which in turn affects their social economic status. However, for some, prevailing opportunities accorded to their refugee status allows them to enjoy exclusive levels of life which at times is at par with the host society.

12.4.4.3 Immigrant Inclusion Level III

This represents the highest level of inclusion enjoyed by migrants whose status is above that of refugees. Migrants who enjoy this status are mainly professionals in the middle- and high-income brackets as well as business entrepreneurs. This is commonly applicable to Europeans, whose racial identity elevates their status to access formal housing, employment and social opportunities. For black people, even at that level, some form of attrition with locals is still observable. This is highlighted in the case of Lusaka where professional immigrants and business people from South Africa, India and China enjoy equality in the city. Similarly, professional immigrants in Johannesburg have access to bank loans which allows them to invest in housing and other facets of luxurious life. But challenges pertaining to language barriers (such as in Maputo where Portuguese is the official language) and job quota systems (in Gaborone and Johannesburg) at times make it difficult to get employment. However, their level of inclusion is not complete because some, depending on their immigration status, cannot vote or partake in age-old pension (e.g. those with permanent residence in South Africa).

In summary, it can be reiterated that achieving inclusivity is not an event – but a protracted and holistic process which hinges on the participation of all stakeholders. In a world where social and economic opportunities are at stake, the recognition of immigrants through formal channels paves the way for inclusiveness. But at individual levels, inclusivity begins with access to economic opportunities. There is no way inclusivity can be achieved if people do not have access to economic opportunities (such as jobs) because this forms the basis upon which access to housing, transport, recreational and other better prospects can be achieved. However, it must also be noted that immigration places a burden on the host country. If it is not handled properly, it negatively impacts on the financial, social and economic status of the host country.

Bibliography

Chipungu, L. & Adebayo, A. A. 2012. The Policy-Planning Divide: An Evaluation of Housing Production in the Aftermath of Operation Murambatsvina in Zimbabwe. *Journal of Housing and the Built Environment*, 28 (2). DOI: 10.1007/s10901-012-9311-8

Chipungu, L. & Magidimisha, H. H. 2021. *Housing in the Aftermath of the Fast Track Land Reform Programme in Zimbabwe*. Routledge.

Government of South Africa. 2021. *Green Paper on Comprehensive Social Security and Retirement Reform*. Pretoria, South Africa: Department of Social Development.

Leshoro, D. 2022. *Government Too Incompetent to Deal with Zama-Zama Cancer*. Johannesburg. South Africa: City Press. 14 August 2022. Available at: https://www.news24.com/citypress/business/government-too-incompetent-to-deal-with-zama-zama-cancer-20220812

Newton, K. 2005. *Foundations of Comparative Politics: Democracies of the Modern World*. Cambridge: Cambridge University Press.

Newton, K. & Van Deth, J. W. 2016. *Foundations of Comparative Politics: Democracies of the Modern World*. Cambridge: Cambridge University Press.

OAU. 1969. *Convention Governing the Specific Aspects of Refugee Problems in Africa*. Addis Ababa, Ethiopia.

UNHCR. 1951. *The Refugee Convention*. Geneva, Switzerland.

UNHCR. 1967. *Protocol Relating to the Status of Refugees*. Geneva, Switzerland.

West, G. 2018. Scale – *The Universal Laws of Life, Growth and Death in Organisms, Cities and Companies*. Amazon Books.

World Bank. 2015a. *A New Approach to Cities: Everyone Counts*. The World Bank.

World Bank. 2015b. *Everyone Counts: Making the Cities of Tomorrow More Inclusive*. The World Bank.

13 The Epilogue

13.1 Introduction

Migration in the Southern African Region is a contentious issue which requires governments to take decisive positions in order to address it. The impact of migrants in recent years has spread into cities as people seek opportunities in diverse sectors. This chapter summarises the discourse of immigrants and inclusive cities by revisiting the key strands that emerged from the discussions. The focus is on the notion of migrants, migrants in the region and inclusive cities. This chapter concludes by providing recommendations which are meant to help policy makers to mitigate against the negativities associated with migration. It also builds on some recommendations provided in Chapter 12.

13.2 Revisiting the Key Strands of the Book

There are three key issues that are at the core of this book – viz. the concept of immigrants, the concept of inclusive cities and the operationalisation of these concepts in the context of Southern Africa.

(a) **The Concept of Immigrants**
   Chapter 1 provided the foundation for this book and the key concepts of immigrants and inclusive cities were discussed to some theoretical depth. What emerges clearly from the discussion is that the concept of immigrants is an omnibus term which is shrouded with diverse connotations which apply to the identity and status of people who decide to move or flee from their place of residence internally or across national borders. Driven by both push and pull factors, the movement of immigrants is at the mercy of the host countries' immigration policies. But further observations show that the success of immigrants largely depends on their immigration status, which is a key determinant of their legality in the host country. Hence the interplay of the status of immigrants and prevailing national policies of the host country (such immigration policies, labour policies which dictate work quotas, etc.) essentially determines their access to opportunities and services provided in the country.

Chapter 2 took this discourse further and looked at the interplay of economics, institutions and assimilation among immigrants.

Unfortunately, most policies tend to favour nationalities of host countries, thereby relegating most immigrants into marginalisation. The situation is aggravated by the status of migrants (especially those who are not documented) who normally fall into *oblivion*. Hence the discourse of immigrants is always punctuated with 'crime and vulnerability' – labels which are associated with the failure by some immigrants to successfully compete for opportunities in host countries, thereby finding themselves engaging in illegal activities or become mere destitutes. However, as shown in most global cities, not all immigrants are vulnerable and not all immigrants are associated with under-dealings that flout rules and regulations of host countries. There are success stories of immigrants who successfully become integrated and assimilated in host communities.

(b) **The Concept of Inclusive Cities**

This is one of the most elusive concepts in contemporary urban studies. As already noted in Chapter 1, where it was discussed to some extensive detail, inclusive cities have both a natural dimension and a human dimension. The natural dimension's focus is on building cities that are sustainable from a natural perspective and whose intention is to protect the natural environment as well as mitigate any consequential negativities (such as those associated with climate change). Man's intervention in the natural environment comes at a cost of depleting what is not easily replaceable. The increase in the city's foot print increasingly impact on the natural environments with wet lands and hilly areas (among others) succumbing to human needs. This environmental concern is a genuine measure of trying to protect the natural environment for mutual benefits with future generations in mind. The sustainable use of urban environments should allow the natural environment to successfully regenerate itself where possible while at the same time be able to support human life. It is the search for this equilibrium which should drive development in urban areas.

The environmental concerns discussed above all stem out of the need to satisfy the human dimension. Building inclusive cities is a function of diverse factors among which are resources, policies and the attitude of the people concerned. The interplay of these factors (i.e. resources, policies and attitude) determine the extent to which cities achieve their mandate of inclusivity. Cities are centres of diversity; and for that reason, any development should strive to reach out to any potential beneficiary in the city. Inclusivity does not necessarily imply that people should get the same level of services – but it revolves around the concept of providing decent services as defined by prevailing international and national policies. In essence, it is a response to achieving the minimum acceptable standards of services as encapsulated in existing policies and within the capacity of the government's resources. Co-production (be it in housing or urban infrastructure), which is yet another entrant in contemporary urban studies concepts, makes it easier to stretch government resources since these are supposed to be boosted by the participation of beneficiaries. This makes the urban poor and other vulnerable groups to be partakers in the development of inclusive cities. However, policy frameworks are human constructs – therefore their success largely depends on the intention of those in power and the attitude of recipients. The success of building inclusive cities therefore rests on implementing policies that shun discrimination and takes cognisance of the needs of the vulnerable too.

(c) **Immigrants and Southern African Cities**

As already noted in Chapter 1, Southern Africa is emerging as one of the key destinations of immigrants from around the world and from within the region. The availability of resources such as mines and oil coupled with highly industrialised cities is attracting both skilled labourers and unskilled labourers. The situation has been aggravated by regional instabilities such as civil wars in Mozambique and DRC Congo, and the land reform programme in Zimbabwe, which generate mass movements of people within and beyond the region. But what is also clear is that immigrants in the region are attracted to those countries where immigration policies are accommodative and where the economy offers opportunities. It is therefore not surprising that Johannesburg, Lusaka and Gaborone are experiencing high numbers of immigrants although restrictive immigration policies, to some extent, tend to deter undocumented immigrants in Zambia and Botswana.

The availability of economic opportunities is the main driver of immigrants in the region. As already noted in the preceding paragraph, diverse economic opportunities in Johannesburg, Lusaka and Gaborone make them preferred destinations by documented immigrants. Once they secure opportunities, documented immigrants find it easy to settle and integrate in these cities. Their ability to support themselves financially enables them to compete successfully with locals for urban services such as housing, educational services and health facilities. For this reason, their level of assimilation in society is much higher. While opportunities do exist in Mozambique, the major drawback is the language barrier (Portuguese) which makes it difficult for immigrants from English-speaking countries to get opportunities and settle. However, most small-scale entrepreneurial immigrants from Africa and abroad still find the country accommodative and they have since established their businesses. Maputo is one city where immigrants do not much complain about discrimination of whatsoever. However, its major drawback is the level of infrastructure and services since Mozambique is among the least developed countries in the world.

Southern African cities are also home to undocumented immigrants. While most of these come from neighbouring countries where they cross porous national borders, some are a result of human trafficking syndicates that disregard national immigration laws. South Africa's immigration laws, which allow visitors to seek asylum after crossing the border, for instance, have seen unprecedented number of immigrants from the region who end up overstaying and eventually reside illegally. This has not only created social upheavals (such as xenophobia) but is making it difficult for cities to provide adequate services. The existing level of polarisation in cities between the 'haves and have nots' does not only point to the legacy of apartheid, but also to challenges arising out of the post-apartheid period associated with high urbanisation. Unfortunately, immigrants are a part of this equation. The situation has been worsened by the long history of cultural links and colonial labour laws which make some illegal immigrants, especially from Lesotho and Swaziland, disregard national borders as mere artificial boundaries. On the other hand, the level of inclusivity in other cities such as Harare, Lusaka and Gaborone for undocumented immigrants is different in the sense that they are restricted to refugee camps. In this regard, undocumented immigrants emerge as vulnerable members of society who are not only recognised, let alone welcomed by society.

13.3 Recommendations

As already noted in Chapter 1, achieving inclusivity in cities is as elusive as the concept of inclusive cities itself. This stems from the fact that there are many factors which should be considered. The situation, in the context of this book, is aggravated by the diversity of immigrants in different geopolitical and economic setups of countries in the region. Nevertheless, an attempt to provide a way forward is discussed in this section. The recommendations proposed herein are meant to be guidelines and the choice of what and how to implement them largely depends on individual cities' priorities as informed and guided by national policies. Recommendations so provided in this section are informed by prevailing conditions of cities discussed in this book and supported by international references drawn from a wide spectrum of expert knowledge provided by international bodies such as the United Nations. In this regard, the criteria taken is to present a holistic approach where both concepts (of inclusivity and immigrants) are discussed together with the intention of creating conducive environments for immigrants in cities of host countries.

(a) Immigrants in a Global Perspective

It should be remembered that migration and immigrants are protected by international laws. The Universal Declaration of Human Rights of 1948 (UN, 2015) has five different articles which relate to immigrants and these are summarised in Table 13.1. This is an important point of departure which elevates immigration from being a mere movement of people between national states. It places immigration at a global level where analysis of immigrants goes beyond counting numbers to incorporate movement of capital, technology, culture and institutional reforms (among others) which are beneficial to the host country. Above all,

Table 13.1 Rights of Immigrants

Article Number	Rights of Immigrants
Article 13	The right to freedom of movement and the right to leave the country of origin and even to return.
Article 14	The right to seek and enjoy asylum in other countries.
Article 23	The right to work.
Article 25	The right to a standard of living adequate for health and well-being … including food, clothing, housing, medical care, social services and security in case of unemployment, sickness, disability, widowhood, old age or any other lack of livelihood.
Article 26	The right to education.
Article 27	The right to freely participate in the cultural life of the host community.

Source: Universal Declaration of Human Rights (UN, 2015).

compartmentalising immigration to national setups discards the relevance of historical dynamics which to a large extent have huge impact on contemporary developments (Portes and Böröcz, 1989).

These rights, to a large extent, form the basis for building inclusive cities because these rights are not a prerogative of immigrants alone, but apply to all humankind. Therefore, the search for key inputs that feed into building inclusive cities is not far-fetched because they are there in principle. These are the same principles which underline most countries' national constitutions. They form part of fundamental building blocks which should be included in the search for achieving the human dimension of inclusive cities. However, the failure to incorporate them into developmental agendas arises out of many factors among which are the availability of resources to invest into these various sectors.

(b) Data for Inclusive Development

One of the biggest challenges that bedevils immigrants and inclusive development at city level is lack of data that can effectively inform development. For a very long time, immigrants have been perceived to fall under the responsibility of national governments whose mandates revolve around enacting immigration policies and monitoring movements (among others). At the city level, immigrants are 'invisible' despite their physical presence. In reality, the impact of immigration is felt at municipal/city level yet basic data that relates to their demographics does not exist. Developing inclusive cities requires disaggregated inclusive data at the city level which captures the number of immigrants (be they legal or illegal), sex, age, disability type, income and geographical location. More so, collecting disaggregated inclusive data is not a once off thing – but a protracted continuous process which local authorities should always conduct. Conducting baselines should not be seen as the sole responsibility of cities, but is a shared responsibility with other sector departments, regional governments and other stakeholders. This in turn forms the basis for understanding barriers that limit inclusion and accessibility to services and opportunities while at the same time it provides the platform for intervention strategies such as master plans, integrated development plans and local development plans (UCLG, 2019).

(c) Creating Immigration Logistical Support Centres

One of the weaknesses experienced at city level is the absence of structures dedicated to immigrants. As already noted in the preceding paragraph, lack of such centres creates a gap in the spatial strategies of the city which arises out of lack of inclusive data. If put in place, such centres can create the basis for collecting data and monitoring movements of immigrants through identifying immigrant entry-points into the city and evaluating their human needs irrespective of their nationalities, religion, gender and occupation. Such centres in turn become *bridges* for thawing relationships between local authorities and immigrants. They assist in removing the stigma of immigrants being 'unwanted elements' in the city and gives them confidence to come forward to communicate their needs to city officials. This can be done in conjunction with other stakeholders such as government departments (e.g. for asylum seekers) and church organisations who feed immigrants on daily basis. The truth is that, at the city level, such structures already exist but they conduct their business in an uncoordinated manner resulting in most immigrants using their own initiatives to survive among strangers. Intermittent interventions that only surface during crisis such as when there are xenophobic attacks and cleansing of crime spots in cities are not sustainable since they happen at the stage when immigrants are already under duress. Immigration structures are meant to create platforms for constructive dialogue, which in turn become entry points for mitigating the negative impact of immigrants (Chipungu and Magidimisha, 2020).

(d) Inclusive Housing

Access to decent housing is not only a problem experienced by immigrants but is a perennial problem which citizens of host cities also experience. This is inevitable given that over 70% of the spatial development of the city revolves around housing. Housing for immigrants comes in three key forms – through settling in informal settlements, renting in both informal settlements and in conventional housing and through purchase by those who can afford. The state of vulnerability of immigrants requires protracted intervention measures whose options are only accorded to citizens at city level (as dictated by national policies). Given restrictive resources that even fail to provide decent housing to citizen indigents, Southern African cities need to embrace principles of self-housing whose resources combine those of national governments, local authorities and the sweat equity of beneficiaries. Such intervention measures, be they through upgrading of informal settlements or relocation of people to green-fields, provide the most sustainable modality to inclusive development since they are dictated by the affordability of beneficiaries in terms of designing the type of infrastructure and defining the level of provision through available resources. The local authority's major task is to ensure access to land and provision of bulk and trunk infrastructure essential for connecting to local infrastructure at household level. Normally, this is a gradual process which does not financially debilitate both the local authority and beneficiaries. The value of such intervention comes in four forms – it improves the physical structure of the house using appropriate materials, ensures access to proper infrastructure, ensures stable tenure arrangements and ensures sustainability since such housing schemes are incorporated into the city's spatial development strategy. Above all, this is not only a strategy for building sustainable cities, but it also forms the basis for building strong bonds among beneficiaries at the community level through trust, support and shared experience. Lastly, self-help housing limits the chances of families being trapped into debts through start-up solutions which are community-based and driven by families. Hence the success of such intervention measures promotes recognition which at times attracts formal financial institutions to engage in community-driven projects. This in turn helps in fulfilling the aspirations of building inclusive cities through both the environmental and human dimensions of inclusivity.

Another alternative housing intervention measure involves optimising on urban density in order to arrest urban sprawl thereby maximising access to physical and social infrastructure existing in the city (Chipungu et al., 2022). This requires a revision of existing zoning regulations in order to accommodate high-rise land-uses and mixed developments. Higher densities are a viable option where land is scarce and expensive, while demand for public housing is high. The option of mixed developments in designated locations is a value capture mechanism that brings economic efficiency to the residential environment. The 'Refugee City' in Palestine and South Africa's robust social housing policy (The National Housing Code, 2009) are all indicators that point to the feasibility of building inclusive cities through high density and mixed-use approach (UN-Habitat, 2018).

Lastly, rental housing is among one of the entry points for immigrants into the housing market. The unfortunate thing is that most immigrants are vulnerable – and therefore, the bulk of their rental housing market (be it in formal and formal housing environments) is predominately unregulated. Hence migrants are trapped in environments where they overcharged irrespective of the deplorable nature of such housing. This calls for measures that are meant not only to improve living conditions of immigrants, but also the need to enact legislative measures that protect the housing rights of the vulnerable tenants. The need is to put in place legal binding measures to meet minimal housing standards and legally binding lease agreements. This, together with the presence of immigrant centres can help to elevate the status of immigrants in terms of their living conditions in cities.

(e) **Generating Inclusive Income for the City**

No venture, at whatsoever level or magnitude is free. Municipal intervention measures are driven by financial resources, therefore, there is need to promote and encourage hybrid economies where micro businesses, small, medium and large businesses can co-exist. This requires tapping into the innovative entrepreneurial initiatives of immigrants which come in various forms such as street vending, retail shops, kiosks, solid waste management and urban agriculture. Promoting these economic activities potentially helps to boost local economic development since such ventures provide income generating opportunities to both immigrants and local residents (UN-Habitat, 2018). There are a number of benefits that emanate from the recognition of such business ventures and these include:

- They contribute to the fiscus of the local authority through taxation.
- They contribute towards self-sustenance of local residents and immigrants through job creation and income generation thereby reducing dependence on government.
- They provide avenues through which local needs and livelihoods of both immigrants and local residents can be assessed.
- They can be used as the basis for building business partnerships.
- They are avenues through which information can be generated to inform demand-driven skills, vocational training, job counselling and financial start-ups.
- They form the basis for regulating the use of public spaces (such as street vending, urban agriculture, dump sites) through provision of licences/permits, communication of rules/regulations and re-organisation of such spaces to create harmony among competitors and other potential users of such spaces.

(UN-Habitat, 2018)

Southern African cities are home to a myriad of such micro-scale economic activities yet lack of recognition and formal intervention channels make the cities lose revenue and opportunities for propagating inclusivity in cities.

(f) Building Positivity Attitudes for Inclusivity

Inclusive cities are not built by managing immigrants – but by working with migrants. One of the biggest problems that contribute to discrimination of immigrants emanates from the negative attitude of host communities and officials who demean immigrants. The existing policy pronunciations which form the basis for respecting human dignity irrespective of nationality or immigration status are not enough. Driven by local authorities and other stakeholders, building inclusive cities is a negotiated process that starts with a positive mind which forms the basis for constructive negotiations (UN-Habitat, 2002). This requires a protracted process of building inroads into communities to assess and understand perceptions and images of host communities which serve as attitudinal barriers to accommodate immigrants (UCLG, 2019).

Community organisations at the neighbourhood level are key points of entry since they enable people to come together. Social cohesion, through events organised by local authorities and other stakeholders (such as awareness campaigns through song and dance festivals), should be promoted at the community level since they help to reduce tension among different groups. More so, existing informal community structures and support mechanisms for immigrants should be recognised and supported. These form conduits through which effective coordination and communication with local authorities and other stakeholders can be propagated for the benefit of the entire community (Asian Development Bank, 2017). It should therefore be borne in mind that changing people's attitude is not a linear process – but a long subtle process that requires a holistic approach by local authorities yet inclusive of both the host community and immigrants. Everyone has a role to play.

(g) Inclusivity as a Process of Deregulation and Reregulation

Urban settings comprise of regulated spaces. In the same vein of resonance, access to the city is bound by regulatory frameworks that affect both the use of space and the behaviour of city residents. Therefore, the process of building inclusive cities is not linear but a process of bundling, unbundling and re-bundling regulations with the sole mandate of achieving satisficing levels which are accommodative and beneficial to both the physical environment and human behaviour (Lindblom, 2018). Policies and legislations as human constructs are bound to fail as they might not respond to contemporary environmental and human dynamics. Constant reviewing of such regulations in line with emerging challenges and new thinking is essential in order to align developments within the framework of inclusive cities. The UN-Habitat (2018) outlines a number of ways through which such a process can be achieved among which are:

- Reviewing existing immigration laws in line with contemporary changes.
- Empowering immigrants to navigate regulatory environments to bolster their business ventures or simply as law-abiding city residents.
- Identifying and demarcating environmentally sensitive areas, hazardous areas and any public spaces would help to protect the environment from unsanctioned occupations by city dwellers (through informal housing) resulting in negative consequences which impact on both the city and its residents.
- Recognising and participating in regional protocols at national level forms the basis of harmonising inclusive development beyond city boundaries. For instance, the failure to ratify the SADC Immigration protocol of 1996 by some member countries is seen as a drawback in achieving inclusivity for immigrants.

(Nshimbi et al., 2018)

13.4 Concluding Remarks

Cities of Southern Africa are diverse – a sense of diversity drawn from the spatial endowment of resources which in essence – and are engines which drive development and change. Developed over years of intermittent use of both international and local capital, these cities have become emblems of success, power, identity, poverty and conflict (among others), yet they continue being centres of attraction for those in search of opportunities. Through migration, Southern African cities embrace diversity that is built through engagement of people from different races and nationalities. It is this diversity, which historically created conflict, and ironically, crafted a platform for harmony and peace as seen today. However, the reality of the matter is that human nature is governed by conflict which at times manifests itself through subtle means. The search for building inclusive cities with immigrants in mind is a conflict in its own way which pits human nature on one hand and the natural environment on the other hand, but complicated by human nature through diversity born out of immigration.

Migration as a natural phenomenon is here to stay. Its resurgence in recent years, especially in the Southern African region, is a pointer to the need to embrace it as both a social and physical construct. But more so, its reconfiguration as a dynamic process, spreading its tentacles into cities, can only be ignored at our own peril. Cities are already bearing the brunt of its dynamics as epitomised through urbanisation and its associated impact. The call to build inclusive cities is therefore not by accident but is in direct response to the unstoppable effect of urbanisation whose impetus of late has been fuelled by migration across national borders. Regardless of shrinking municipal resources in the face of mounting problems within their jurisdictions, the need to build inclusive cities cannot wait given environmental and human threats cities are already enduring. In the same space of resonance, it will be an anomaly to ignore the impact of immigrants in our societies. The call, in this regard, is to embrace immigrants as partners in the continual journey in search of building inclusive cities. However, balancing the influx of migrants and the needs of citizens in the country is a delicate balance which pits governments as its people, especially where there are societal challenges. President Ramaphosa acknowledged the problem and noted (Matiwane, 2022):

> That balance has to be there for us the first prize is to look after our own people and make sure that their rights are not ignore – but at the same time those who come here (must) have their rights respected.

This book contributes towards building inclusive cities with immigrants in mind. Driven by both empirical evidence and selective secondary data sources, the myriad recommendations provided in this book are a step towards ameliorating the pervasive impact of urbanisation, partly driven by both internal and external migration. The authors are cognisant of the fact that the recommendations provided in this chapter serve as mere guidelines for designing intervention measures that contribute towards building inclusive cities. As such, these recommendations are meant to steer discussions towards building inclusive environments as informed by the context of cities. Detailed coverage of these measured can be further pursued in various sources of information cited by authors in the compilation of this book. The inclusion of three competing concepts (i.e. inclusive cities, immigrants and Southern Africa) could not allow the authors to go into deep discussion of each concept without compromising the other. However, a platform for creating connectivity among these three concepts was created upon which insights into pertinent issues that affect cities of Southern Africa can be evaluated.

Bibliography

Asian Development Bank. 2017. *Enabling Inclusive Cities. A Tool Kit for Urban Development*. Asian Development Bank. Available at: https://doi.org/10.22617/TIM157428

Chipungu, L., Kamuzhanje, J., Makonese, E. D. & Magidimisha-Chipungu, H. H. 2022. Leapfrog Developments–Gaps, Challenges and Opportunities in Urban Development: Cases of Harare and Durban. *Journal of Urban Systems and Innovations for Resilience in Zimbabwe-Jusirz*, 4, 45–65.

Chipungu, L. & Magidimisha, H. H. 2020. *Housing in the Aftermath of the Fast Track Land Reform Programme in Zimbabwe*, Routledge.

Lindblom, C. 2018. The Science of "Muddling Through". In Stein, J. M. (Eds.), *Classic Readings In Urban Planning*, 31–40. Routledge.

Matiwane, Z. 2022. Ramaphosa Says Mec's Foreign Rant Could have been Handled Differently. *Times Live*.

Nshimbi, C. C., Moyo, I. & Gumbo, T. 2018. Between Neoliberal Orthodoxy and Securitisation: Prospects and Challenges for a Borderless Southern African Community. In Magidimisha, H. H., et al. (Eds.), *Crisis, Identity and Migration in Post-Colonial Southern Africa*, 167–186.Springer International Publishing AG.

Portes, A. & Böröcz, J. 1989. Contemporary Immigration: Theoretical Perspectives on its Determinants and Modes of Incorporation. *International Migration Review*, 23, 606–630.

The National Housing Code. 2009. *Social And Rental Interventions. Social Housing Policy*. Vol. 6. South Africa: Department of Human Settlements.

UN-Habitat. 2018. *Tracking Progress Towards Inclusive, Safe, Resilient and Sustainable Cities and Human Settlements*. Un-Habitat. Available at: https://unhabitat.org/review-on-sdg11-synthesis-report-for-the-2018-hlpf-tracking-progress-towards-inclusive-safe

Index

Acculturation–The Four Models Approach 45; Assimilation 46; Separation 46; Integration 46; Marginalisation 46

Balancing act 200

Capital cities 24
Cities as economic organisms 47
Cities as regulated platforms 47
Conditionalities 200

Dimensions of Inclusivity in Cities 6

eGoli 114–135; Constitution of the Republic of South Africa 1996 124; Refugee Act 130 of 1988 125; Immigration Act 13 of 2002 as amended in 2014 127

Immigrants and the City 41–49; second-class citizens 43
Impact of the Pre-colonial Migrations 52–53; State Formation 52; Labour Immigrants 52; The Class System 53
In the Shadows of the Sunshine City of Zimbabwe 153–171; the Citizenship of Zimbabwe Act No. 23 of 1984 (as amended by Act No. 7 of 1990, Act No. 12 of 2001, Act No. 22 of 2001, Act No. 23 of 2001, Act No. 1 of 2002 and Act No. 12 of 2003) 163; the 2013 (or new) Constitution 163; The Refugees Act (Act 13/1983, 22/2001, (s.4)) 165; the 1951 Convention 162
Inclusion of Foreign Migrants in South African Cities 173–192;

Inclusiveness in Spatial Integration 173; Freedom of Access to Public Areas and Services 177; Freedom to Access Service Centres and Amenities 178; Institutional Support and Immigration Policies 186; Role of Media in Xenophobia 188
Inclusive Cities 23
Inclusivity 24
Inscribed into the web of criminals 200
Invisibility 200
Immigrants 24

Lusaka 136–152; The Immigration and Deportation Act No.18 of 2010 145; The Refugee Act of 2017 146; Constitution of Zambia Act No.1 of 2016 146; The Zambian Refugees Control Act of 1970 147

Malawi 80–97; The Malawi Immigration Act 1964 89; The Malawi Refugees Act of 1989 89; Employment Act No. 6 of 2000 90
Maputo 98–112; Mozambique: Act No. 21/91 of 31 December 1991 (Refugee Act) 106; Mozambique: Immigration Law No. 5/93 of 1993 (aliens) 108; Labour Act 109
Migrant Profile in Gaborone 66
Migrant Profile in Lilongwe 82
Migrant Profile in Maputo 100
Migrants Profile in Johannesburg 117
Migrants Profile in Lusaka, Zambia 140
Migrants' Profile in Harare 154
Migration Laws 194; Legal immigrants 194; Illegal migrants 194

Protection fees 200

Research Methodology 20–29

Social cohesion and community integration 13
Southern African Countries 24

The Anatomy of Inclusivity in Cities 2
The Anchors of Inclusive Cities 4
The human dimension of cities 47
The Post-colonial Protectionist City 60
The Right to the City 2
The Sustainable Development Goals 3
Typology of Immigrants **9**

Unpacking Migrant Laws in Gaborone 64–77; The Refugee (Recognition and Control) Act 1967 72; Botswana's Constitution of 1966 with Amendments through 2016 73; The Geneva Convention Act 75

Who Are Immigrants in Cities? 8; An immigrant as foreign-born 8; An immigrant as a foreigner or non-national 8

Zama-zama 200

Printed in the United States
by Baker & Taylor Publisher Services